Digital Radiography

Selected Topics

DIGITAL RADIOGRAPHY
Selected Topics

Digital Radiography

Selected Topics

Edited by

James G. Kereiakes
Stephen R. Thomas

University of Cincinnati College of Medicine
Cincinnati, Ohio

and

Colin G. Orton

Wayne State University School of Medicine
Harper-Grace Hospitals
Detroit, Michigan

Plenum Press • New York and London

Library of Congress Cataloging in Publication Data

Digital radiography.

Includes bibliographies and index.
1. Radiography, Medical—Digital techniques. 2. Diagnostic imaging—Data processing. 3. Image processing—Digital techniques. I. Kereiakes, James G., 1924- . II. Thomas, Stephen R. III. Orton, Colin G. [DNLM: 1. Radiology—instrumentation. 2. Radiography—methods. WN 200 D5745]
RC78.7.D35D54 1986 616.07′57 86-3259

ISBN-13: 978-1-4684-5070-5 e-ISBN-13: 978-1-4684-5068-2
DOI: 10.1007/978-1-4684-5068-2

A Division of Plenum Publishing Corporation
233 Spring Street, New York, N.Y. 10013

Softcover reprint of the hardcover 1st edition 1986

To our wives

Helen, Ingrid, and Barbara

for their continuing support and encouragement

Contributors

R. E. Alvarez DigiRad Corporation, Palo Alto, California 94303

B. A. Arnold Image Analysis, Inc., Irvine, California 92714, and Department of Radiological Sciences, UCLA Medical School, Los Angeles, California 90024

R. Brennecke Medical Clinic, Johannes Gutenberg University, D-6500 Mainz, Federal Republic of Germany

W. L. Henry Department of Medicine (Cardiology), University of California, Irvine, California 92716

J. G. Kereiakes Department of Radiology, University of Cincinnati College of Medicine, Cincinnati, Ohio 45267

R. A. Kruger Department of Radiology, University of Utah College of Medicine, Salt Lake City, Utah 84132

A. V. Lando Department of Radiological Sciences, University of California, Irvine, California 92716

L. A. Lehmann DigiRad Corporation, Palo Alto, California 94303

O. Nalcioglu Department of Radiological Sciences, University of California, Irvine, California 92716

S. J. Riederer Department of Radiology, Duke University Medical Center, Durham, North Carolina 27710

W. W. Roeck Department of Radiological Sciences, University of California, Irvine, California 92716

J. A. Seibert Department of Radiology, University of California, Davis, California 95817

S. R. Thomas Department of Radiology, University of Cincinnati College of Medicine, Cincinnati, Ohio 45267

J. M. Tobis Department of Medicine (Cardiology), University of California, Irvine, California 92716

Preface

Digital radiography is a general term describing any projection radiological system in which the image exists in digital form at some stage between acquisition and viewing. In an earlier form, radiographic films were digitized in an attempt to enhance and redisplay information of interest. The field has evolved to its current state, in which X-ray signals are detected electronically, converted to digital form, and processed prior to being recorded and displayed. A primary goal of digital radiography is the removal of interfering effects from secondary structures in an image, so that clinically significant details can be displayed with enhanced visibility. The achievement of this goal involves many parameters, including contrast agents, subtraction techniques, processing techniques, filtering techniques, system noise, and quantitative aspects.

It is the purpose of this book to present material by noted individuals in the field covering several of the above topics.

The authors acknowledge the secretarial and editorial assistance of Mrs. Helen Taylor and the editorial assistance of Mrs. Ruth McDevitt.

James G. Kereiakes
Stephen R. Thomas
Cincinnati, Ohio

Colin G. Orton
Detroit, Michigan

Contents

1. *DIGITAL RADIOGRAPHY: OVERVIEW*

B. A. Arnold, J. G. Kereiakes, and S. R. Thomas

2. *IMAGE PROCESSORS FOR DIGITAL ANGIOGRAPHY: ALGORITHMS AND ARCHITECTURES*

R. Brennecke

3. TEMPORAL INTEGRATION PROCESSING TECHNIQUES

S. J. Riederer

4. NOISE ANALYSIS IN DIGITAL RADIOGRAPHY

B. A. Arnold

5. QUANTITATIVE ASPECTS OF IMAGE INTENSIFIER–TELEVISION-BASED DIGITAL X-RAY IMAGING

O. Nalcioglu, W. W. Roeck, J. A. Seibert, A. V. Lando, J. M. Tobis, and W. L. Henry

6. RECURSIVE FILTERING TECHNIQUES APPLIED TO DIGITAL SUBTRACTION ANGIOGRAPHY

R. A. Kruger

7. ENERGY-SELECTIVE RADIOGRAPHY: A REVIEW

L. A. Lehmann and R. E. Alvarez

1

Digital Radiography

Overview

B. A. Arnold, J. G. Kereiakes, and S. R. Thomas

1. INTRODUCTION

Digital radiography has been used loosely to refer to an entire class of X-ray imaging systems, frequently without reference to the image detector used or the image processing functions applied. These systems have in common the acquisition of a two-dimensional projection image in digital format but include a variety of detectors, techniques, and applications. The key attributes of these systems are improvements in detector design and the ability to process the image (in some cases in real time) to greatly improve the capability for image information extraction. Several image processing functions have been developed which include image subtractions, image additions and averaging, contrast enhancement and windowing, point transformations, correction for detector nonlinear response, and others. The ability to process the image rapidly and in digital format has allowed time mode image subtraction to be carried out conveniently as well as permitting contrast enhancement of the subtracted image. Fairly comprehensive reviews of digital radiography can be found in recent publications by Price *et al.*,[1] Kruger and Riederer,[2] Brody,[3] and Arnold.[4,5]

Digital radiography systems can be classified into three groups based on the form of the primary X-ray detector and the image acquisition procedure: point-scanned detectors, line-scanned detectors, and area de-

B. A. Arnold • Image Analysis, Inc., Irvine, California 92714, and Department of Radiological Sciences, UCLA Medical School, Los Angeles, California 90024. **J. G. Kereiakes and S. R. Thomas** • Department of Radiology, University of Cincinnati College of Medicine, Cincinnati, Ohio 45267.

TABLE 1. Detector Systems for Digital Radiography[5]

Type	Detector	Advantages	Disadvantages
Point-scanned	NaI scintillator photomultiplier tube	High quantum detection efficiency Excellent scatter removal Very low electronic noise Beam area determines spatial resolution Single detector avoids spatial sensitivity problems	Very inefficient use of radiation Tube power loading limitations Long exposure times High quantum noise Poor spatial resolution Low frame rate
Line-scanned	Phosphor screen–Si photodiode array CT-type xenon ionization chamber	Low-noise, wide-dynamic-range detector system Good scatter removal Medium to high quantum detection efficiency "Focal plane shutter effect" avoids motion blur High contrast detectability	Inefficient use of radiation Tube heat load limitations Exposure time, noise, or resolution compromised Low frame rate (limiting time mode subtraction) Limited spatial resolution (to maintain high quantum efficiency) Variations in detector-to-detector sensitivity
Area	Selenium photoconductor-microelectrometers Stimulable phosphor-scanning laser photodetector CsI image-intensifier TV camera	Very efficient use of X-ray exposure Good spatial resolution potential Good contrast sensitivity potential Medium to high quantum detection efficiency Medium to high frame rates (allowing dynamic recordings with video systems)	Poor scatter removal Readout device may limit resolution and/or add noise Specific for image intensifier: Limited dynamic range of TV camera Limited field of view Potential for additive noise source in the video chain

tectors. These systems all have in common the acquisition of a two-dimensional projection image. Upon closer inspection it becomes apparent that some systems may contain a combination of two approaches. For example, a detector may be exposed as an area detector while being read out on a point-by-point basis or by a line-scan system. Table 1 lists some detectors used in each type of imaging system.

Digital X-ray images acquired from densitometric scans of X-ray films were evaluated for several medical applications in the 1960s and could be considered the first digital radiographs. Digital imaging processing of medical X-ray images at that time appears to have failed because of the limitations of the primary detector, X-ray film. It is important to note that the success of these new classes of digital imaging systems is as dependent on the development of X-ray detectors as it is on digital computers.

Figure 1 indicates schematically the various detector systems which have been applied in digital radiography.[6]

2. POINT-SCANNED DETECTOR SYSTEMS

Point-scan radiography utilizes a high-sensitivity, stationary detector system from which images are generated by scanning a pencil beam of radiation over the patient.[7,8] The technique consists of measuring the intensity of radiation transmitted through the patient by a pencil X-ray beam which is scanned in a raster pattern over the body area to be imaged.

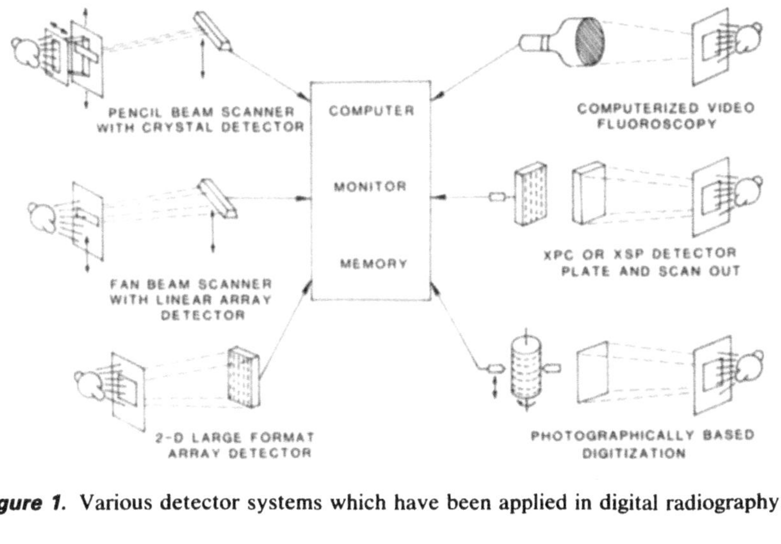

Figure 1. Various detector systems which have been applied in digital radiography.

The X-ray detector consists of a relatively large NaI scintillation crystal and photomultiplier tube. The detector is sufficiently large to cover the entire width of the X-ray beam formed by the collimator. Successive lines of the image are generated by translating the X-ray tube, collimator, and detector with respect to the patient. The output from the detector is digitized and input to a computer for storage and image processing.

The advantages and disadvantages of point-scan detector systems are listed in Table 1. Point-scan radiographic systems have two major disadvantages. First, the frame rate is low compared to area or line-scan detectors due to the long exposure times required to generate the images. Second, and more important, the maximum heat load of current X-ray tubes limits the total photon fluence which can be delivered to the patient in a given exposure time. This limitation cannot be overcome, since the image is created by scanning in a raster fashion and is therefore created one pixel at a time. For example, each pixel is exposed for a very small fraction [1/(480 × 512)] of the total beam "on" time. At present, this corresponds to an effective exposure time of 0.06 msec as compared to approximately 100 msec with area detectors. A major breakthrough in X-ray tube technology would be necessary to achieve the desired photon statistics necessary for most diagnostic applications.

3. LINE-SCANNED DETECTOR SYSTEMS

Line-scanned radiographic systems utilize a linear array of detectors which are scanned across the patient to acquire a two-dimensional projection image. The use of X-ray intensifying screens in combination with photodiode arrays has been reported by Sashin[9] and Mattson and Barnes,[10] and the evaluation of xenon ionization detectors as used in CT scanners has been reported by Brody[11] and Cohen.[12]

A major limitation of line-scan systems is the inefficient utilization of X-ray tube power loading. Acquisition of the image on a line-by-line basis limits the total photon fluence which can be delivered to the patient in a given time period. This may represent a major limitation of the scan detector approach since photon statistics play a major limiting role in image quality. A second related limitation is the time required to move the detector mechanism across the patient to acquire the image. Long scan times create the potential for patient movement which may preclude time mode subtractions in most applications.

Line-scan detector systems which utilize discrete detectors have an additional limitation due to the desire to simultaneously increase spatial resolution and quantum detection efficiency (QDE). In general, as the detector units become smaller and achieve higher resolution, the QDE is also reduced. This is especially true in the case of xenon detectors since the spacer between detector cells becomes large relative to the detector sensitive volume. Detector-to-detector variation in sensitivity is another problem of line scanners which requires correction in the computer to achieve high contrast detectability.

4. AREA DETECTOR SYSTEMS

Area detector systems overcome the primary limitation of line-scan and point-scan systems in that a full two-dimensional image is achieved by flash X-ray exposure of the detector system. Three substantially different area detector systems are presently available or are under development, as indicated in Table 1. The selenium photoconductor detector is read out by an array of scanning microelectrometers. It is therefore an area detector in the sense of exposure and a line-scan system in terms of image readout. The newest digital radiography detector system under development utilizes a stimulable phosphor which is read out by a scanning laser beam in combination with a photodetector. This system is an area detector in the sense of exposure, but is read out on a point-by-point basis. The third area detector system, which utilizes an image intensifier in combination with a television camera, has received the most commercial focus to date due to the clinical success of digital video subtraction angiography. These detector systems will be discussed in greater detail below.

4.1. Stimulable Phosphors

A recent area detector system under development for digital radiography utilizes a stimulable phosphor which is read out by a scanning laser beam. Table 2 lists some of the properties of the stimulable phosphor detector system. This detector system has two distinct stages of image generation, the exposure and the readout stages which are required to be separated in time. A plate containing the stimulable phosphor is exposed to the X-ray beam much like a screen–film cassette. A latent image is generated by storage of the absorbed X-ray energy in trapping sites within the phosphor. The latent image is then released as fluorescent light following stimulation of the phosphor by visible or ultraviolet light. Stimulation is implemented by a scanner laser beam of an appropriate wavelength. The latent image is then reproduced by detection of the emitted fluorescent light by a photodetector and converted to an electrical signal

TABLE 2. Properties of Stimulable Phosphor X-Ray Imaging Detectors[5]

X-ray energy is stored in trapping sites
Stimulation by visible or UV light releases stored energy
Emitted light is detected by a photodetector
Storage time of hours
Wavelength of emitted light ≤ 500 nm

for digitization. The wavelength of the fluorescent light must be sufficiently different from that of the laser beam to allow separation of the light sources before detection in the photomultiplier tube. Stimulable phosphor thicknesses of 0.3 mm provide high QDE since medium- to high-*Z* materials make up the phosphor. The time required for image readout and the intensity of the light required for stimulation are major determinants of spatial resolution since the laser beam can be made very small in diameter. Readout times of 1 to 3 min are expected, placing it in a range comparable to present film development.

Stimulable phosphors contain alkaline earth metal fluorohalides consisting of at least one divalent metal and at least one halogen. One element selected from the group of Eu, Tb, Ce, Tm, Dy, Pr, Ho, and Er is also required. The availability of sufficient light emission from the phosphor and the performance capabilities of presently available photomultiplier tubes can be expected to lead to high signal-to-noise ratio (SNR) images primarily limited by X-ray quantum statistics. The major concerns of this system at present appear to be image storage and readout times at desired levels of spatial resolution.

4.2. Selenium Detectors

Amorphous selenium has been known for some time to be an excellent detector of X-rays.[13] It is well known as a detector in xeroradiography systems for mammography. In this case the thickness of the plate is about 50 to 100 μm which is adequate at the relatively low energies used. Efforts have been directed toward developing a digital radiography system for general radiography at higher photon energies using selenium as the basic detector. A major problem has been the scarcity of selenium plates with sufficient thicknesses to produce good QDEs. The selenium detector plates can be read out in several ways. The conventional method uses powder cloud development as in xeroradiography systems, while newer approaches utilize either a laser readout or linear arrays of microelectrometers. Digital selenium-based systems have the potential for high spatial resolution as well as high QDE. Development of improved detector plates and appropriate readout devices, which are practical in the clinical environment, will be required for commercial success.

4.3. Digital Video Systems

Digital video radiography, the most commercially advanced of the digital radiography systems, utilizes a more or less conventional X-ray image intensifier in combination with a high-quality TV camera. Digital video subtraction angiography has emerged rapidly during recent years following early developmental work at several institutions.[14–19] During the early 1970s, analog video subtraction techniques were investigated by Mistretta *et al.*[20] using storage tube techniques. It was shown that low-contrast iodine signals could be isolated by time and energy subtraction techniques using image-intensifier TV systems. Image degradation in the analog devices limited the applications of these techniques. During the late 1970s, digital video image processing systems were constructed at the University of Wisconsin, the University of Arizona, and by the Quantex Corporation. Two- and three-spectrum *K*-edge and TID subtraction procedures were developed by Mistretta and co-workers at the University of Wisconsin.[21,22] Improvements in the design and optimization of video imaging chains were achieved during the same time period. Clinical feasibility studies using digital subtraction techniques for intravenous angiography examinations began in 1979.[23–25] Developmental programs in digital video radiography began at South Bay Hospital[26] and large-scale clinical trials were carried out at the Cleveland Clinic.[27]

Digital intravenous (IV) angiography has been successfully applied throughout most of the body. Following the clinical success of IV examinations, it soon became apparent that digital imaging systems and associated image processing techniques could be used to advantage for interarterial studies as well. Improvements and extensions of these techniques are continuing to be developed. Worthy objectives remaining include not only the development of new clinical procedures and improved image quality for more accurate diagnosis but also improved patient safety through reduced radiation and dye dosages, improved patient comfort, and increased clinical efficiency. Investigational efforts continue in digital angiocardiography toward the goal of improving the visualization of coronary vessels, bypass grafts, and cardiac chambers, and to provide quantitative techniques for cardiac function analysis.[28]

The development of improved X-ray detectors and fast digital processing has accelerated interest in digital imaging systems for other radiographic applications. The image-intensifier TV camera has been shown to be an excellent imaging detector for subtraction angiography. Arnold has investigated the potential of digital video radiography, with and without subtraction, for applications in gastrointestinal and intravenous pyelogram examinations.[29]

Optimization of the digital video radiography system has led to total-system SNRs of 700:1 for subtraction imaging. The image quality achieved in the final image depends heavily on the performance of the imaging chain,[30] and the requirements will vary with the type of examination; e.g., carotid IV studies versus cardiac studies. The major advantage of video systems is the ability to acquire images rapidly at rates up to 30 video frames/sec, allowing time mode subtractions to be implemented as well as the capability of recording dynamic information. Limitations of the video approach are the restricted dynamic range of the TV camera and the potential for additive noise sources in the video chain.

The advantages and disadvantages of area detectors are listed in Table 1. A key advantage of area detectors is the very efficient utilization of X-ray tube power loading, leading to short exposure times and the potential for high frame rates. The detector materials utilized permit good spatial resolution and high QDEs. Very high frame rates (see Table 3) can be achieved with the video system, while frame rates of 2 to 3 images/sec appear reasonable for the selenium detector and the stimulable phosphor approaches. All three area detector systems share the common disadvantage of poor scatter removal properties of the two-dimensional detector. This requires the use of scatter reduction methods such as the grid, air gap, or possibly, in the future, digital scatter reduction techniques.[31]

5. COMPARISON OF X-RAY IMAGING SYSTEMS

It is now generally recognized that conventional screen–film radiography results in the loss of a great deal of diagnostic information contained in the exit beam from the patient. The CT scanner has demonstrated that many soft tissue structures can now be visualized which were previously not seen using conventional radiography. It is useful to compare the performance capabilities of screen–film combinations with those of present digital radiography systems (Table 3). The values listed should be considered approximate since the measurements and calculations are representative of a variety of systems and exposure conditions. The spatial resolutions listed in Table 3 are representative of limiting values for high-contrast objects. Contrast detectability is defined as the minimum detectable subject contrast for larger, nonresolution-limited details (5-mm diameter or larger). The contrast detectability values given for the digital

TABLE 3. Performance Specifications of X-Ray Imaging Systems[5]

	Spatial resolution (lp/mm)	Contrast detectability	SNR (max.)	Patient dose (rad)	Images/sec
Screen–film radiography	4–8	2%	200 : 1	0.2	2
Digital radiography					
Video	2	0.5%	700 : 1	0.2	30
Line-scan	1–2	0.2%	500 : 1	0.07	0.5
Point-scan	1–2	2%	170 : 1	0.004	0.2

radiography systems are representative of those achievable with time mode subtraction in iodine contrast medium studies. The SNRs are determined from the maximum signal divided by the rms noise of this signal. The values given are approximate and dependent on detector electronic noise and dynamic range as well as quantum noise. The SNRs are representative of image areas equal to a pixel size in all cases except for the screen–film system where a 0.5-mm-diameter area was used. The radiation dose levels are poorly standardized and probably represent the greatest uncertainty of all the parameters listed. For example, the thickness of the object and beam energies are not standardized. The maximum image repetition rates are listed in the final column. The values given for the line-scanned systems are based on expected scan times of machines presently being developed and are not directly related to the maximum SNRs listed.

The major advantages of screen–film systems are high spatial resolution and the convenient archival storage afforded by film. These are important advantages which may be difficult to overcome in routine radiological practice. Digital radiography systems provide two-dimensional projection images while providing moderate spatial resolution and good contrast detectability. All three digital radiography systems achieve comparable spatial resolutions but differ significantly in terms of their minimum contrast detectability and the number of images per second which can be obtained.

The difference in contrast detectability between the point- and line-scan systems can be explained on the basis of photon statistics, i.e., the dose is only 4 mrad for the point-scan system. The line-scan systems have excellent low-contrast detectability (5-mm-diameter objects) at exposure levels which can be achieved with existing X-ray tubes. The necessary requirements to increase the kVp to provide higher photon flux levels reduce image contrast which may be an important limitation in some examinations. The most distinct advantage of the video systems is their high frame rate which cannot be matched by any of the other imaging systems. It is this high frame rate and the efficient utilization of X-ray tube power loading that has allowed digital video subtraction systems to be used for time mode subtraction studies in angiography.

6. SUMMARY

Digital subtraction IV angiography is a currently accepted clinical modality. Quantitative angiocardiography is developing rapidly and promises to be a valuable clinical procedure with widespread application. Digital video radiography and beam energy subtraction barium studies warrant further technical and clinical development. Digital electronic radiography is a promising new technology which has important implications in the evolution from film-based radiography to digital electronic medical imaging.

REFERENCES

1. R. R. Price, F. D. Rollo, W. E. Monahan, and A. E. James, *Digital Radiography: A Focus on Clinical Utility,* Grune & Stratton, New York (1982).
2. R. A. Kruger and S. J. Riederer, *Basic Concepts of Digital Subtraction Angiography,* Hall, Boston (1984).
3. W. R. Brody, *Digital Radiography,* Raven Press, New York (1984).
4. B. Arnold, H. Eisenberg, D. Borger, and A. Metherell, Digital radiography: An overview, *Proc. SPIE* **273,** 215–226 (1981).
5. B. A. Arnold, Digital radiography: A technology overview, *Proc. SPIE* **347,** 7–14 (1982).
6. Adapted from K. Vizy, *Overview of Digital Angiography,* Eastman Kodak Company (1981).
7. Y. Tateno and H. Tanka, Low dosage x-ray imaging system employing flying spot x-ray microbeam (dynamic scanner), *Radiology* **121,** 189–195 (1976).
8. P. J. Bjorkholm, E. Annis, and E. Frederick, Digital radiography, *Proc. SPIE* **233,** 137–144 (1980).
9. D. Sashin, E. J. Sternglass, M. J. Spisak, J. Boyer, K. Bron, L. Davis, D. Gur, J. Herron, R. Hoy, W. Kennedy, and K. Preston, Computer electronic radiography for early detection of vascular disease, *Proc. SPIE* **173,** 88–96 (1979).
10. R. A. Mattson, R. A. Sones, J. B. Stickney, Q. Barnes, and M. M. Tesic, Design and physical characteristics of a digital chest unit, *Proc. SPIE* **314,** 160–163 (1981).
11. W. R. Brody, A. Macovski, and L. Lehmann, Intravenous angiography using scanned projection radiography: Preliminary investigation of a new method, *Invest. Radiol.* **15,** 220–223 (1980).
12. G. Cohen, L. K. Wagner, S. R. Amtey, and F. A. DiBianca, Contrast-detail dose and dose efficiency analysis of a scanned digital and a screen–film–grid radiographic system, *Med. Phys.* **8,** 358–367 (1981).

13. L. S. Jeromin and L. M. Klynn, Electronic recording of x-ray images, *J. Appl. Photogr. Eng. SPSE* **5**(4), 183–189 (1979).
14. R. Brennecke, T. K. Brown, J. Bursch, and P. H. Heintzen, Computerized video-image processing with application to cardioangiographic roentgen image series, in *Digital Image Processing* (H. H. Nigel, ed.), Springer, Berlin (1977).
15. B. K. Gilbert, M. T. Storma, et al., A real time hardware system for digital processing of wide band video signals, *IEEE Trans. Comput.* **C-25**, 1089–1100 (1976).
16. M. M. Frost, H. D. Fisher, S. Nudelman, and H. Roehing, A digital video acquisition system for extraction of subvisual information in diagnostic medical imaging, *Proc. SPIE* **127**, 208–215 (1977).
17. R. A. Kruger, C. A. Mistretta, J. Lancaster, et al., A digital video image processor for real time x-ray subtraction imaging, *Opt. Eng.* **17**, 652–657 (1978).
18. H. Roehrig, M. Frost, R. Baker, et al., High resolution low level video systems for diagnostic radiology, *Proc. SPIE* **78**, 102 (1976).
19. R. A. Kruger, C. A. Mistretta, T. L. Houk, et al., Computerized fluoroscopy in real time for noninvasive visualization of the cardiovascular system: Preliminary studies, *Radiology* **130**, 49–57 (1979).
20. C. A. Mistretta, M. G. Ort, J. R. Cameron, et al., Multiple image subtraction technique for enhancing low contrast periodic objects, *Invest. Radiol.* **8**, 43–44 (1973).
21. R. A. Kruger, C. A. Mistretta, and A. B. Crummy, Digital k-edge subtraction radiography, *Radiology* **125**, 243–245 (1977).
22. F. Kelcz and C. A. Mistretta, Absorption edge fluoroscopy using a 3-spectrum technique, *Med. Phys.* **3**, 159–168 (1976).
23. C. M. Strother, J. F. Sackett, A. B. Crummy, F. G. Lilleas, C. A. Mistretta, R. Kruger, D. Ergun, and C. G. Shaw, Clinical applications of computerized fluoroscopy: The extracranial carotid artery, *Radiology* **136**, 781–783 (1980).
24. T. W. Ovitt, P. C. Christenson, H. D. Fisher, M. M. Frost, S. Nudelman, H. Roehrig, and G. Seeley, Intravenous angiography using digital video subtraction: X-ray imaging system, *Am. J. Roentgenol.* **135**, 1411–1414 (1980).
25. B. A. Arnold, H. Eisenberg, D. Borger, et al., Digital video subtraction angiographic imaging, in *Radiological Society of North America, Annual Meeting,* Dallas (1980).
26. B. A. Arnold, H. Eisenberg, D. Borger et al., Real time digital video subtraction x-ray imaging, *Scientific Exhibit, Radiological Society of North America, Annual Meeting,* Dallas (1980).
27. M. A. Weinstein, W. A. Chilcote, M. E. T. Modic, T. P. Meaney, et al., Digital subtraction carotid angiography: A comparative study with conventional angiography in 100 patients, in *Radiological Society of North America, Annual Meeting,* Dallas (1980).
28. W. Barrett, H. Eisenberg, B. Arnold, and P. Scheibe, High speed processing of digital intravenous angiocardiography images for enhancement of coronary bypass grafts and quantitation of left ventricular function, in *Computers in Cardiology, Proceedings IEEE,* Seattle (1982).
29. B. A. Arnold, P. Scheibe, H. Eisenberg, et al., Digital video radiography: Applications in GI and IVP examinations, *Proc. SPIE* **347**, 278–285 (1982).
30. B. A. Arnold, H. Eisenberg, D. Borger, et al., Digital video subtraction angiography, in *Proceedings of the Fifteenth International Congress of Radiology* (A. Wackenheim, ed.), pp. 162–176 (1981).
31. J. Boone, A scatter correction algorithm for digitally acquired radiographs (SCADAR), Ph.D. thesis, University of California, Irvine (1985).

2

Image Processors for Digital Angiography

Algorithms and Architectures

R. Brennecke

1. INTRODUCTION

After a period of experimental and clinical development,[1–9] digital processing of angiographic X-ray video image sequences is now routinely applied in clinical and research work. The clinical advantages offered by this approach have been discussed in several reports.[10–12] The primary application is the improved visualization of regions of the heart and circulation opacified by X-ray contrast material during angiographic and angiocardiographic examinations. More complex techniques are being developed for improved functional analysis based on digitized angiograms. Technically, the digital techniques also potentially offer improved means of acquiring, storing, and handling images when compared to present film-based angiography. It is therefore expected that in many applications the film will eventually be replaced by digital photoelectronic methods of X-ray image acquisition, archiving, and communication.[9,13,14]

The extremely high data rate and data capacity required by angiographic, and especially by angiocardiographic imaging, preclude the use of conventional general-purpose data processing systems for the purposes mentioned above. Typical special-purpose processor systems for image analysis are optimized for enhancement and feature extraction of static

Most of this work was prepared while the author was with the Department of Pediatric Cardiology and Biomedical Engineering, University of Kiel, Federal Republic of Germany.

R. Brennecke • Medical Clinic, Johannes Gutenberg University, D-6500 Mainz, Federal Republic of Germany.

images by two-dimensional filters, while algorithms for digital angiography operate primarily on dynamic image sequences and therefore require a much higher input/output activity. The integration of the system resources needed for real-time image series acquisition with those needed for fast image sequence processing has to be one of the main goals of system design in digital angiography.

In response to this need, in recent years a number of special-purpose processors have been developed by university groups and by industry. After a short discussion of typical algorithms applied in digital angiography, we will review the basic design goals and architectures developed. It is important to note that no survey of machines is intended and that, instead, a coverage of the basic principles of present designs is given.

2. ALGORITHMS FOR HANDLING AND PROCESSING OF DIGITIZED ANGIOGRAMS

Detailed reviews of qualitative image enhancement[15,16] and image analysis techniques[17] developed for digital angiography have been published previously. Here, only a short characterization of the algorithms developed is given.

2.1. Image Data Compression

Some early systems for digital angiography handled simple tasks by real-time digital processing and subsequent analog storage of enhanced (e.g., subtracted) image sequences. This can be accompanied by irreversible loss of information. Moreover, complex tasks can only be handled by storage of digitized image data and subsequent off-line processing. However, standard mass storage devices such as computer disks or tapes are not compatible with the high data rate required, especially by angiocardiographic image sequences (40 to 240 Mbit/sec).

Techniques for data compression are known from numerous applications[18] in image transmission but they are usually not information preserving. Deviations of the displayed image from the original one are tolerated if they are not conspicuous. Much experience will be needed to find out which degree of error can be tolerated in medical imaging used for diagnostic purposes. A more conservative approach to data compression is, therefore, the application of information-preserving (reversible) data compression. We have shown that simple intraframe prediction and Huffman coding can lead to appreciable reductions in data rates and data capacity requirements.[19] Similar results were obtained by Lehmann and Macovski using adaptive predictors.[20]

Irrelevance reduction is another, self-evident technique for data compression. The definition of the relevance of data is primarily a medical problem. Digital systems offer in principle the flexibility to use any asynchronous imaging rate, e.g., when triggering image acquisition by the asynchronous ECG-signal.[2] Moreover, real-time filtering of noisy image data may in some cases compress data without loss of essential information. At the same time, many of the techniques applied in irrelevance reduction also result in a reduction in X-ray dose. This is an aspect that deserves further intense study.

2.2. Image Enhancement by Digital Subtraction

The basic image processing technique applied in digital angiography is temporal image subtraction. Upon closer inspection, it is apparent that this method requires the execution of several steps including temporal low-pass filtering (for noise reduction), temporal image subtraction (for extraction of angiographic details from the background), and contrast enhancement by linear rescaling or histogram modification techniques[1,21]

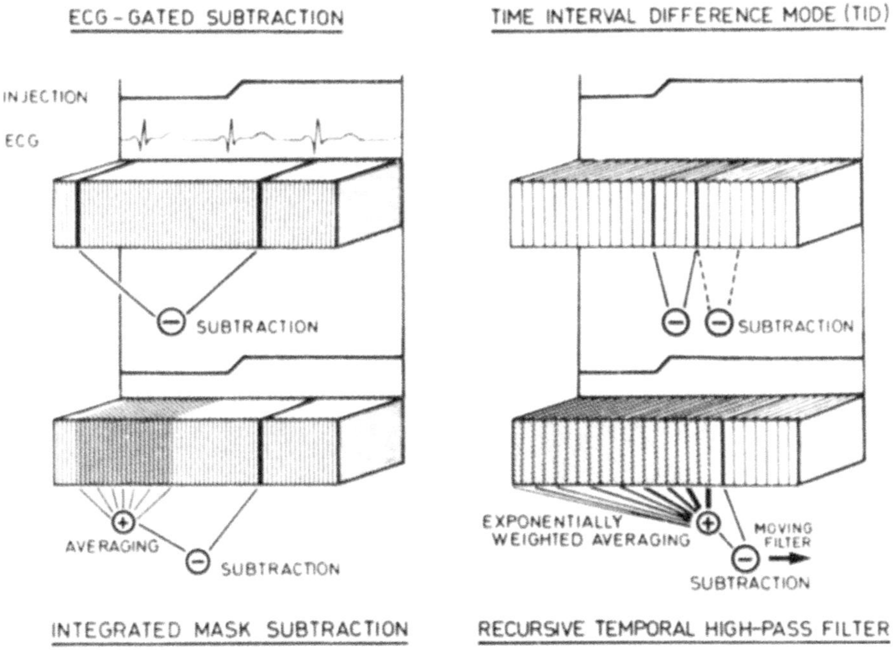

Figure 1. Schematic representation of four basic subtraction techniques for the enhancement of angiocardiographic image sequences.

in order to match the signal amplitudes to the range of gray scale values best visualized by the observer.

Figure 1 gives an overview of typical subtraction techniques developed in recent years. Low-pass filtering can be by convolutional or by recursive algorithms. The task to select well-matched background images for subtraction from a given dye image can make the seemingly simple subtraction process quite complex. An approach to automatic selection of suitable background images is based on two-dimensional regional similarity detection by cross-correlation.[19] An alternative approach, applicable when only small movements occur, is the rotation or warping of the background image for optimum match.[22,23]

2.3. Image Enhancement and Extraction by Digital Filtering of Pixeldensograms

Figure 2 shows schematically the time courses of the gray level density for two pixels positioned over a vessel. We call these time function pixeldensograms. A generalized approach for angiographic image sequence processing regards digital filtering of these pixeldensograms (64,000 to 256,000 in a typical image sequence) as the basis for image sequence enhancement and analysis. Typical algorithms applied include linear and nonlinear filters, such as recursive band-pass[19,24,25] and median filtering, as well as many feature extraction methods based on earlier experience in X-ray densitometry.[26–29] Correlation filters have been shown to be an efficient tool in these applications for amplitude and time-parameter extraction.[30,31] It is predicted, however, that in the future increased use

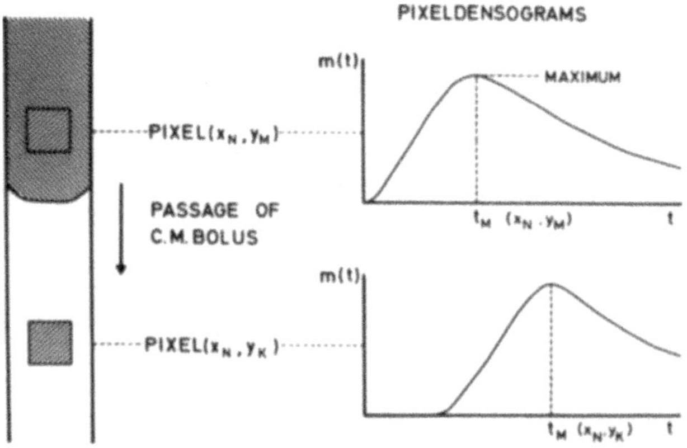

Figure 2. Principle of time-parameter extraction from image sequences using local analysis. The density–time curves measured in each pixel ("pixeldensogram") are evaluated separately to determine parameters such as the time of maximum opacification (right).

will be made of more general-context information contained in the neighborhood of each pixel extending into the spatial and temporal (three-dimensional) vicinity.[17]

2.4. Image Analysis

Quantitative diagnostic information on physiological parameters, such as heart border movement or vascular blood flow, can be derived by measurements based on time-varying distances[12] and/or densities.[17] The preprocessing steps described above often improve these measurements. Distance measurements are still performed mainly by manual outlining of the structures of interest, while densitometric measurements require typical image processing techniques such as segmentation of an image into regions with certain properties.[29]

2.5. Algorithm Structures

2.5.1. Point Operations

Point operations are typically needed in logarithmic transformations[21] performed before subtraction, and for linear and nonlinear rescaling of subtraction image data.[1] Look-up tables are the standard means of achieving these transformations.

2.5.2. Filtering of Pixeldensograms

Filtering of pixeldensograms is the most often used approach to image sequence enhancement and parameter extraction. In principle each of the 64,000 to 256,000 pixeldensograms (time-series) is processed separately. Therefore, in the following discussion the spatial coordinates of the pixels are omitted. The algorithms include subtraction from a constant:

$$\ln y_i = \ln x_i - \ln x_0 \tag{1}$$

convolution filtering:

$$y_i = \sum_{i-n_0}^{i+n_0} k_n x_n \tag{2}$$

and recursive filtering. A second-order recursive filter[30] is described by

$$y_i = k_A y_{n-1} + k_B y_{n-2} + k_C x_{n-2} + k_D x_{n-1} + x_n \tag{3}$$

where y are output samples, x are input samples, and the k's are weighting coefficients (real values). The selection of these coefficients depends on the angiograms to be filtered. Brennecke *et al.*[30] and Kruger *et al.*[31] have described techniques for matching filters to the contrast material bolus dynamics. The basic element of the described algorithms is the multiply/accumulate operation.

In feature extraction, the simplest operator extracts the gray level corresponding to maximum opacification for each pixeldensogram:

$$y = \mathrm{MAX}\,(x_i) = x_m \tag{4}$$

while in time-parameter extraction the time of occurrence of this opacification is determined by

$$y \sim m \tag{5}$$

where m is the index of the sample fulfilling Eq. (4). Böhm and Höhne[32] and Brennecke *et al.*[30] discussed several more complex time-parameters. All of the described operations do not require global program branching operations, a fact greatly simplifying their realization.

2.5.3. Two-Dimensional Processing

Two-dimensional filters are used mainly for image registration (see Section 2.2). If the area over which the similarity analysis is performed is limited to 10% or less of the total image, then "brute force" implementation using fast multipliers and accumulators as the basic elements is applicable.

Another two-dimensional filter problem is the restoration of scatter effects. These problems belong to the class of the inverse filters.[33] Shaw *et al.*[34] applied straightforward averaging over a large neighborhood as an approximation of the convolution operation, but filters with more detailed weighting coefficient matrices will probably be required. Here, fast Fourier techniques should be preferred because of their much faster performance in large filters.

2.6. Data Structure

The usual way of organizing image information is in rectangular-shaped arrays of vertical dimension A ("lines") and horizontal dimension B ("pixels per line"). Each element of this matrix corresponds to the gray level of a pixel digitized to D bits. Products A * B between 256^2 and 1024^2 have been used. While flexible partitioning of a large physical memory for different image matrix sizes is not difficult, the choice of depth D usually requires large "quantum jumps" in word length. The word length is typically 8 to 10 bit in digitization and 16 bit in most computations on the "pixel" level. Image sequences are acquired and stored as a series of two-dimensional arrays of size A × B and word length D (Figure 3A). However, as discussed in Section 2.3, the pixeldensogram is often the primary data structure analyzed in image sequence processing (Figure 3B). Memory designs that support fast access to pixeldensograms—and not only to images—are to be preferred. This will probably favor large semiconductor memories even when fast disk storage is available. Often, three levels of memory hierarchy are detectable. In the mass memory, sequential files (stored during the acquisition phase) can be read with good efficiency, while this "granularity" becomes finer in the main memory, and random access to pixel information is provided by the functional units performing computations.

Temporal and spatial masks or "windows" marking selected images and regions of interest play an important role in image sequence processing. In principle, spatial masks can be arranged as image planes in much the same way as graphic overlays. A more efficient way of masking is selective access by "intelligent" memory address controllers if cubic windows (two spatial and one temporal dimension) can be used.

A third layer of image information has to be provided for graphics and/or text displayed together with the angiographic image information.

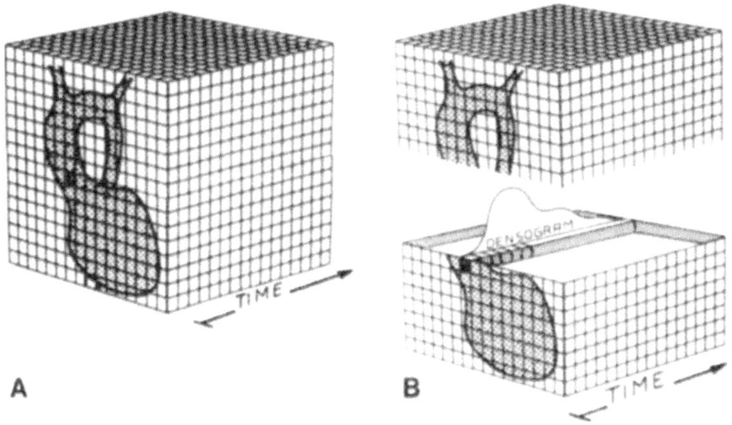

Figure 3. (A) Storage and (B) processing of image sequences in typical algorithms applied in digital angiography. While data are stored as a time-sequence of two-dimensional data arrays, they should be accessible for data processing as vectors (pixeldensograms; see Figure 2). Adapted from Heintzen *et al.*[51]

3. PROCESSOR ARCHITECTURES FOR DIGITAL ANGIOGRAPHY

General-purpose computers offer maximum flexibility in performing virtually every computational and control task. However, the high-speed input–output operations required by digital video processing cannot be handled by these systems. Moreover, the computation times required for image sequence processing prohibit the routine clinical application of image enhancement and analysis techniques. At the other extreme, hard-wired processors can be fast enough but they offer only limited flexibility. In this section we will describe differing system architectures developed to tackle some or all of these problems. Less detailed reviews covering the general field of medical image processing systems have been published by Alexander,[23] Ameling,[46] and in a book edited by Onoe *et al.*[47]

3.1. General-Purpose Computer with Video Interface

Among the first digital image processing systems applied in digital video angiography were those using a standard minicomputer system for data processing, with a special bidirectional video–computer interface for the adaptation of the high data rate of the video system [analog-to-digital conversion (ADC), display] to the much slower data paths of the computer. The earliest video–computer interfaces used analog media (video tape or disk) as fast buffers. Digitization was performed during subsequent slow-motion replay.[35] Another approach used was digitally freezing selected frames in real time in the normal replay mode of a tape recorder.[1] Subsequently, the stored data were transferred to the computer at a much slower rate. All of these techniques have the disadvantages of a limited signal-to-noise ratio (SNR) determined by the video tape recorder, of time base jitter, and of limited spatial resolution.

Real-time digitization is limited to a rate of a few fields per second when using standard digital disk memories with their mean data rate of 5 to 20 Mbit/sec. Ovitt *et al.*[5] described primary storage on fast digital tape (up to 240 Mbit/sec) and subsequent slow-motion transfer of data to computer disk. Controllers using several disks in parallel, and disk systems using several read/write heads in parallel are under development.[36] Thus, the additional need for a special tape system will be eliminated.

The logical next step is the storage of image data in a semiconductor memory with dual-port access to the input–output system and to the computer. An early example of this architecture was described by Gilbert *et al.*[37] This concept will probably become economically attractive even for angiocardiography with its large storage requirements when mass production of dynamic semiconductor memories reaches the Mbit/chip limit.

The principal problem of all of these systems is the limitation of the speed of data processing inherent in conventional single-CPU sequential computers. Therefore, their main application at present is in basic research into the development and evaluation of new algorithms, where the unlimited flexibility of these general-purpose computers and the ease of programming in a high-level language play the most important role.

3.2. Special-Purpose Processors for Real-Time Subtraction

The improvement in conspicuity of angiographic details provided by subtraction is so important, that some of the first digital processors developed were designed to handle just this application out of the numerous processing methods described in Section 3.1. Brennecke *et al.*[1,2] designed a simple system of this type (Figure 4). During image generation, a single background image was stored in a digital field memory. All subsequent frames were subtracted from this information, and the resulting subtraction image sequence was recorded on videotape. The stored mask image was appended to this recording after digital-to-analog conversion.

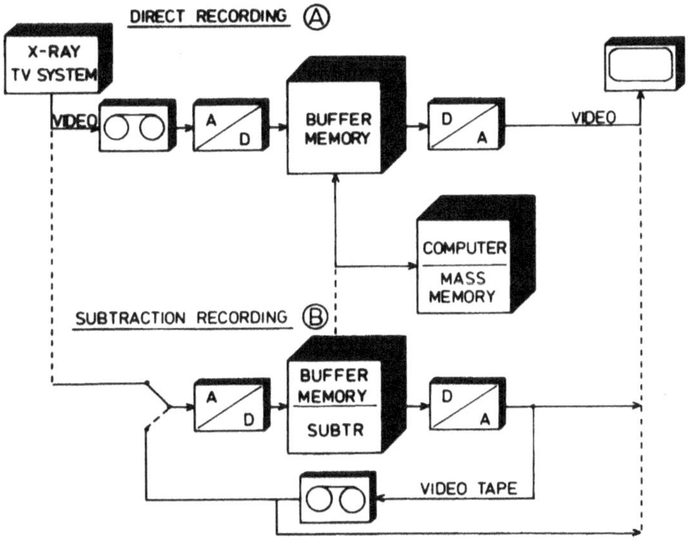

Figure 4. Comparison of a conventional image processing system (A), a general-purpose computer with video interface with (B), a real-time subtraction processor. In (B), the difference between a mask-image (buffer memory) and the life image sequence is computed and amplified before storage on tape. These data may be postprocessed by the subtraction processor or by the general-purpose computer (dashed line in B).

The primary advantage of this technique is the instant availability of enhanced images. A secondary advantage is the gain in SNR provided by subtraction recording. The relevant information of the video angiogram, i.e., the intensity change caused by temporal changes in X-ray absorption, is amplified before recording and the ratio of this specific signal to the noise inherent in (analog) video tape recording is increased accordingly. In the scheme described, background information, which is removed from the recording in the subtraction process, can be added during postprocessing if this information is appended to the subtraction scene. Results obtained by postprocessing using a minicomputer have been reported.[(2)]

The real-time videoprocessor described by Mengers[(38)] had a similar structure but in addition, it offered first-order recursive filtering for noise reduction. It was not developed primarily for medical applications but for low-light-level vision systems. Other early work in digital video processing centered on noise reduction by motion-dependent filtering[(39)] for use in electronic news gathering.

A special-purpose processor for angiography described by Kruger *et al.*[(40)] used frame integration instead of recursive filtering for noise reduction. Subtraction recording was also described.[(6)] In addition, these workers implemented time-interval differencing, a simple temporal high-pass filter accentuating moving structures (Figure 1). This procedure required the use of two memories instead of the single image memory of the other designs. The commercial DVI 1 system manufactured by Philips has a similar structure.[(41)]

Figure 5 compares the structure of standard computers and these special-purpose video processors. These processors have in common that their application is limited to certain types of subtraction. The subtraction results obtained in real time can often be improved by postprocessing. However, when larger patient movements occur, information can be lost irreversibly in the recording process, since the amplified subtraction signal may be clipped by overloading. Finally, the spatial resolution is limited by the bandwidth of the analog (video) recorder to a value much lower than in film recording. The concept is certainly still viable for small compact systems, e.g., for critical care units, and it can be extended by adding conventional image analysis systems for ventriculography.

The SNR of data storage and the flexibility of processing can be improved by adding a real-time digital image recorder[(42,43)] to the special-purpose processor. Peppler *et al.*[(36)] described the DVIP II, a system consisting of a video digitizer (ADC), three semiconductor memories that can be configured to match various image sizes between $256 \times 256 \times 6$ and $1024 \times 1024 \times 12$ bits, and a real-time digital video disk recorder.[(42)] Real-time processors for image integration and subtraction are configur-

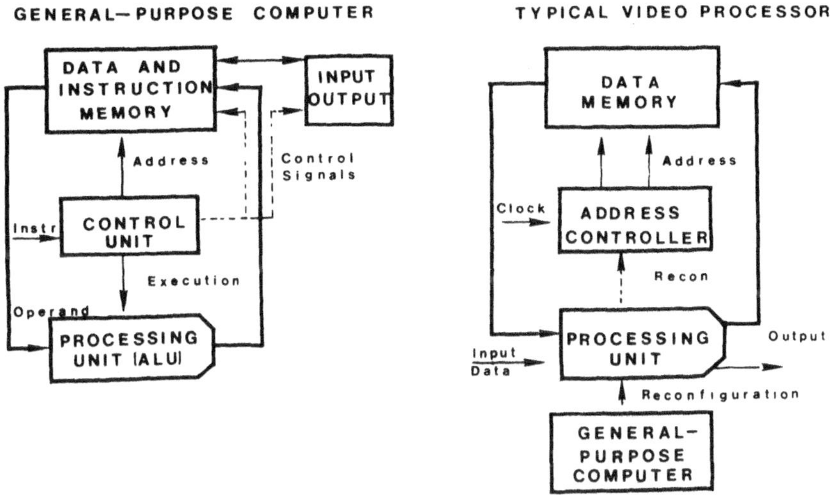

Figure 5. Comparison of the structure of the general-purpose computer (left: von Neumann architecture) and a typical video processor for subtraction imaging (right). The video processor shown is not programmable but it may be reconfigurable for different tasks such as subtraction or TID imaging (see Figure 1). Reconfiguration is performed by the small general-purpose computer (microcomputer).

able by a programmable control computer. Logarithmic processing and scatter correction are performed by additional hard-wired functional units. Applications including temporal and energy subtraction have been reported.[36]

3.3. *Special Computer Systems for Digital Angiography*

While limited computational speed was the main disadvantage of the systems described in Section 3.1, the specialization to subtraction imaging was the principal limitation of the concepts discussed in Section 3.2. In both cases, large investments in memory systems, ADC and DAC modules, and support circuits such as power supplies, were necessary to obtain even limited performance. However, there are improved architectures offering both speed and flexibility of operation.

These architectures can best be deduced from the well-known concept of the array processor attached to a standard minicomputer. An array processor (not to be confused with an array of processors) is attached to a standard general-purpose computer to free it from repetitive purely computational tasks (vector and array processing). The host computer handles all input–output and supervisory tasks. Programs and data are loaded down to the array processor which, in turn, provides processed data and status information (Figure 6A). However, this concept does not work in image sequence processing, since the input–output tasks cannot be handled by the common host computers and the passage of data from mass memory via the main memory of the host to the array processor produces a severe bottleneck. Complex bus systems for 16- and 32-bit microprocessors may reduce this problem in the future.[44] For the present, the changed array processor concept shown in Figure 6B solves the problem by coupling the input–output ports to the attached processor and not to the host computer. Thus, the host computer handles mainly supervisory tasks. In addition, in some cases, the host computer will perform complex computational problems after data reduction by preprocessing handled by the attached processor in the manner described.

The main features of this architecture can be summarized as follows:

1. The tasks to be performed are distributed between a general-purpose host computer (HC) and at least one special-purpose image sequence processor (ISP).
2. The ISP handles digitized video data acquisition, storage, and display (image I/O operations), and high-speed computational tasks

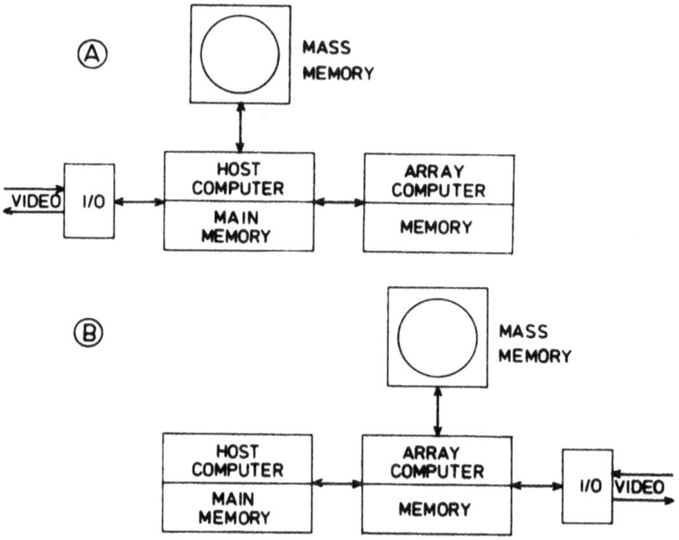

Figure 6. Conventional (A) and modified (B) array processor configuration. The modified scheme reduces the problems of data transfer between host computer and array processor that can become a bottleneck in conventional systems, especially in digital video processing.

of repetitive character (low frequency of global branching operations). Examples are real-time image subtraction, image rotation, and fast matched filtering.
3. The HC downloads programs and commands to the ISP, and receives massively reduced image data and status information from the ISP. Complex operations in pattern recognition or image segmentation are then performed by the HC.
4. The HC will supervise future image management and communication systems.[45]

This scheme, consisting of HC and ISP, is often extended to incorporate small microprocessor systems interfacing with physiological signals of the patient, with the X-ray system, and with the user. We call this the microprocessor controller level (MCL).

The definitions given above are not sufficient to differentiate in an unambiguous way between the special-purpose processors described in Section 3.2 and the more flexible special-purpose computers discussed here. In fact, there is an area of overlap between these concepts. Following is a list of design criteria important in classifying angiographic image processing systems:

1. Flexible systems need a much closer coupling between HC and ISP than the typical subtraction processors.
2. In flexible systems, the timing of the storage and computational operations in the ISP is usually decoupled from the timing of the image acquisition process (analog-to-digital conversion). In the other systems, the integration of asynchronous devices (including the data transfer to the host) is difficult.
3. Flexible systems usually apply time-multiplexed interconnection structures linking I/O units, memories, and computational units. Specialized systems use spatial multiplexing instead. The dynamic reconfiguration of time-multiplexed (bus-oriented) systems and their expansion are generally easier than in spatially multiplexed systems. The principal structure typical for the latter group of processors is shown in Figure 7. An example of a time-multiplexed system is given in Figure 9.

It is a primary problem of algorithm design to match the algorithms to a processor structure that is economically feasible. Figure 8 gives an overview of the main tasks to be performed by an angiographic processor system. While complexity of the algorithms to be performed increases from left to right, the amount of data decreases in the same direction.

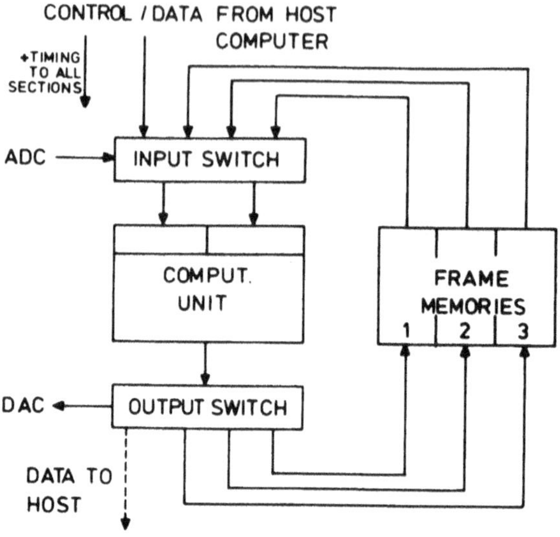

Figure 7. Typical architecture of a conventional image processor. Spatial multiplexing using dedicated data paths instead of time-multiplexed bus systems (see Figure 9) complicates the dynamic reconfiguration of the system and its modular expansion. Central timing synchronized by external processes (e.g., analog-to-digital conversion) complicates the interfacing of asynchronous components such as magnetic disk drives.

Thus, the processes shown at the left and in the middle will be handled primarily by the attached processor, while tasks shown at the right will be performed by the HC.

At least two commercial systems for digital angiography implement the configuration shown in Figure 6A/B by coupling a standard array processor (AP) to a standard minicomputer system (HC). This straightforward approach, however, is not expected to represent the typical system architecture of the future. APs are usually not optimized for fast video input–output requirements (including disk storage). Their theoretical high speed is often not achievable due to overhead factors between the HC and the AP because the match is not optimal for some algorithms between AP architecture and the structure of the image processing problem. The integration of a large (dynamic RAM) frame memory into the structure of the AP can slow down its operation. Finally, the high precision of floating point operations typically offered by APs is not often required in digital angiography, so that the corresponding algorithms make little or no use of this expensive feature of the AP. In the next section, we will therefore discuss two ISPs specially designed for digital angiography.

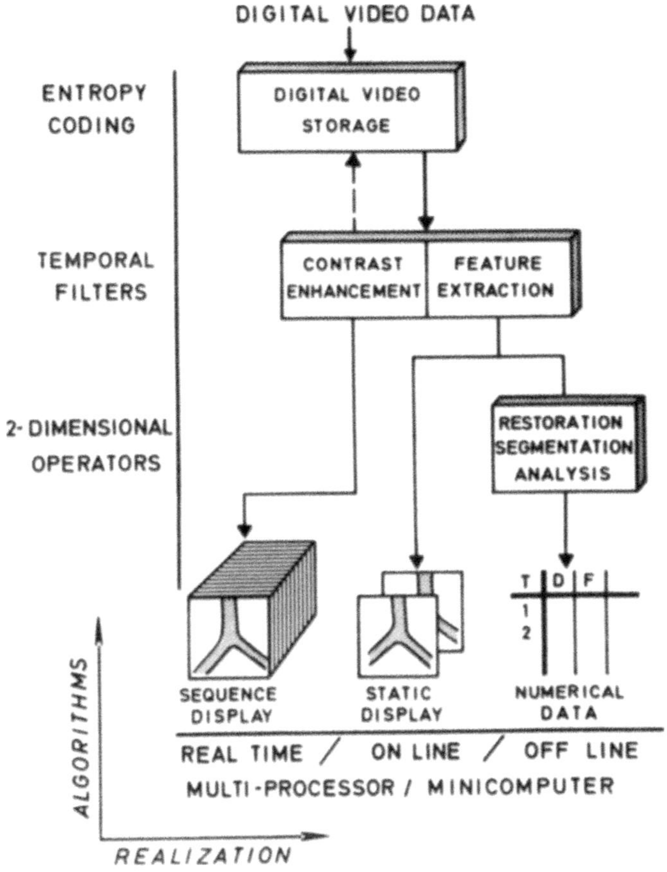

Figure 8. Overview on algorithms for image sequence processing in digital angiography. While the complexity of the algorithms increases from left to right, the amount of data to be processed will usually be reduced. This structure of computational problems can be mapped to the structure of image processor systems consisting of special-purpose and general-purpose computers.

3.4. Experimental Systems for Digital Angiography

The limitations of systems based on the combination of standard APs with general-purpose computers led to the development of computers optimized for image sequence acquisition and processing. Following is a detailed description of two architectures specifically developed for digital angiography. There are certainly other designs that could also be adapted to this task.[13,14,46]

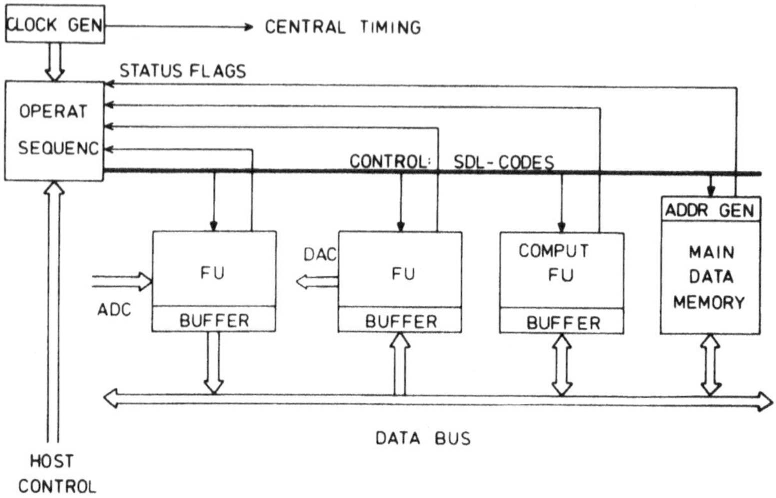

Figure 9. Bus structure of an image sequence processor. The operation sequencer controls access to the time-multiplexed data bus by constituting source/destination links (SDL) between functional units (FU) and main data memory. The frequency of access for a given FU is determined by a control list programmed using the host computer.

Both systems described here use time-multiplexed bus systems for interconnection of the digital video input–output devices, the memory devices, the special-purpose computational units [functional units (FU)], and the HC. This time-multiplexed design provides for standard interfaces with all devices on the bus. Modular expansion is possible as long as the bus bandwidth is sufficiently high to support parallel operations. The bandwidth of the systems described is 30 and 40 Mbyte/sec, respectively. A second characteristic of both systems is the high semiconductor memory capacity (8 and 16 Mbyte).

The DVP system (Hamburg University and DESY) described by Nicolae and Höhne[48] couples a DEC PDP-11 computer to an ISP. The interconnection scheme in the ISP is a fast asynchronous bus. The bus is used as a shared communication path between processors, I/O units, and memory, and it manages the process synchronization according to priorities defined by the user. A bus controller (arbiter) with administration times of 60 nsec was designed to give the processor possessing highest priority fast access to the bus.

The ISAAC system (Kiel University) described by Brennecke *et al.*[49] consists also of a PDP-11 computer (host) and the system components enumerated above connected by a central bus. However, the time-critical and often inflexible process of asynchronous bus arbitration (bus access granting) is eliminated by a programmed scheme. The special synchronous bus (Figure 9) developed for the ISAAC system is described below.

Access to the bus is controlled by an operation sequencer. Each operation is initiated by applying the addresses of a data source and a data sink to the bus. This address pair is called a source/destination link (SDL-code). Similarly, a main memory partition is accessed by specifying the address of the corresponding main memory address generator as either the source or the sink in an SDL-code, while the address of an FU is specified as the other SDL-code in the address pair. These SDL-codes are compared by each FU and each of the main memory address generators to its own address. During the subsequent cycle of the bus clock, data are transferred between the data buffer of the addressed FU or the main memory partition using the data bus. Concurrently, a new SDL-code is transmitted to all FUs. The data transfer initializes the operation of the addressed FU. While these operations proceed, data that are addressed by subsequent SDL-code pairs may be transferred between other FUs.

The bus communication scheme described was developed to give the programmer of the system detailed control over bus access. The sequence of SDL-codes necessary for a given operation is programmed into a list read in a cyclical way. The following processes can be optimized by programming:

1. Allocation of bus time to the data exchange between FUs and memory because of prior knowledge of specific requirements of each unit in a given process. No hardware arbitration is required.
2. Set up of interleaving schemes for memory partitions. Interleaving is used for speed up in large memory systems composed of relatively slow dynamic semiconductor memories. Address pipelining not described here is added to make virtually invisible the access time of the memory components.
3. Translation of the programmed sequence of SDL-codes into structured data streams. This structure of each data stream simplifies the design of FUs by reducing the amount of local storage required and by simplifying address generation on the level of the FUs.
4. Synchronization of two or more processes by making the execution of an SDL-code dependent on the prior execution of a certain other code. This feature, called SDL-code chaining, guarantees the coordination of parallel and pipelined tasks without the need of passing flags between the FUs.

The ISAAC system can be regarded as a multiprocessor synchronized by data streams, while in conventional designs the access of the processors to the data is the primary event. This philosophy, coupled with a data-driven design[50] of the computational and I/O units, avoids some of the problems of latency times in memory access and of deadlocks occurring in more conventional designs due to competition of the functional units for common resources, such as global memory and the bus structure.

4. CONCLUSIONS AND DISCUSSION

Digital systems for angiographic image processing are used to store, retrieve, handle, and process unusually large amounts of data. Both the large amount of data involved and the high speed requirements of the input–output operations of data handling and of data analysis lead to the evolution of special architectures of image processors for this new field. General-purpose architectures fulfilling the requirements stated above would be prohibitively expensive. The special architectures have to be matched to successful algorithms often developed by using slower general-purpose systems. Besides image subtraction, image registration and parametric imaging modes become increasingly important. Those architectures that apply regular time-multiplexed bus structures and a large central memory shared by several processes will probably predominate since they offer the advantage of fast dynamic reconfiguration and modular expandability. Even more important, these structures should simplify the adaptation of a given processor structure to differing applications in areas of medical image processing such as digital angiocardiography, echocardiography, nuclear medical imaging, and nuclear magnetic resonance imaging. This makes conceivable families of medical image processors sharing architecture, physical modules, and system software. A prerequisite for this development is the definition of module interface standards accepted by the industry as a whole.[52] Another problem not discussed in this hardware- and algorithm-oriented review is the need for advanced tools for software engineering and improved design of user interfaces[53] matching the flexibility provided by future hardware architectures.

ACKNOWLEDGMENTS. Many stimulating discussions with Professor P. Heintzen, Kiel University, are gratefully acknowledged. The cooperation of J. H. Bürsch, H. J. Hahne, M. Lüdtke, and W. Petersen contributed to the work described here. This work was supported by grants from the Deutsche Forschungsgemeinschaft.

REFERENCES

1. R. Brennecke, T. K. Brown, J. H. Bürsch, and P. H. Heintzen, Digital processing of videoangiographic image series using a minicomputer, in *Proc. Comput., Cardiol.*, pp. 255–260 (1976).
2. R. Brennecke, T. K. Brown, J. H. Bürsch, and P. H. Heintzen, Computerized video-image preprocessing with applications to cardio-angiographic roentgen-image series, in *Digital Image Processing* (H. H. Nagel, ed.), Springer, Berlin, pp. 244–262 (1977).
3. P. H. Heintzen, R. Brennecke, and J. H. Bürsch, Computer quantitation of angiocardiographic images, *Proc. SPIE* **167,** 17–21 (1978).
4. R. A. Kruger, C. A. Mistretta, T. L. Houk, S. J. Riederer, and C. G. Shaw, Computerized fluoroscopy in real time for noninvasive visualization of the cardiovascular system, *Radiology* **130,** 49–57 (1979).
5. T. W. Ovitt, P. C. Christenson, H. D. Fisher, M. M. Frost, S. Nudelman, H. Roehrig, and G. Seeley, Intravenous angiography using digital video subtraction: The X-ray imaging system, *Am. J. Roentgenol.* **135,** 1141–1144 (1980).
6. A. B. Crummy, C. M. Strother, J. F. Sackett, D. L. Ergun, C. G. Shaw, R. A. Kruger, C. A. Mistretta, W. D. Turnipseed, R. P. Lieberman, P. D. Myerowitz, and F. F. Ruzicka, Computerized fluoroscopy: Digital subtraction for intravenous angiocardiography and arteriography, *Am. J. Roentgenol.* **135,** 1131–1140 (1980).
7. T. F. Meaney, M. A. Weinstein, E. Buonocore, W. Pavlicek, G. P. Borkowski, J. E. Gallagher, B. Sufka, and W. J. MacIntyre, Digital subtraction angiography of the human cardiovascular system, *Am. J. Roentgenol.* **135,** 1153–1160 (1980).
8. R. Brennecke, J. H. Bürsch, H. G. Bogren, and P. H. Heintzen, Digital intravenous imaging techniques in pediatric cardiology, in *Digital Subtraction Arteriography* (C. A. Mistretta, A. B. Crummy, C. M. Strother, and J. F. Jackett, eds.), Year Book, Chicago, pp. 133–141 (1982).
9. S. Nudelman, Photoelectronic-digital radiology, in *Digital Imaging in Cardiovascular Radiology* (P. H. Heintzen and R. Brennecke, eds.), Thieme, Stuttgart, pp. 1–14 (1983).
10. C. A. Mistretta, A. B. Crummy, C. M. Strother, and J. F. Jackett (eds.), *Digital Subtraction Arteriography,* Year Book, Chicago (1982).
11. P. H. Heintzen and R. Brennecke (eds.), *Digital Imaging in Cardiovascular Radiology,* Thieme, Stuttgart (1983).
12. O. Ratib and W. Rutishauser, Recent developments in cardiac digital radiography, *Int. J. Cardiac Imaging,* 29–40 (1985).
13. A. J. Duerinckx (ed.), *Picture Archiving and Communications Systems, Proc. SPIE* **318** (1982).
14. R. H. Schneider and S. J. Dwyer III (eds.), Applications of optical instrumentation in medicine XIII, *Proc. SPIE* 535 (1985).
15. C. A. Mistretta, Digital videoangiography, *Diagn. Imag.* **3,** 14–18, (1981).
16. R. Brennecke, Digital processing of video image sequences, in *Digital Imaging in Cardiovascular Radiology* (P. H. Heintzen and R. Brennecke, eds.), Thieme, Stuttgart, pp. 24–34 (1983).
17. R. Brennecke and J. H. Bürsch, Functional analysis of angiograms by digital image processing techniques, in *Selected Topics in Image Science* (O. Nalcioglu and Z. H. Chow, eds.), Springer, Berlin, pp. 182–216 (1984).
18. A. N. Netravali and J. O. Limb, Picture coding: A review, *Proc. IEEE* **68,** 366–406 (1980).
19. R. Brennecke, J. H. Hahne, K. Moldenhauer, J. H. Bürsch, and P. H. Heintzen,

Improved digital read-time processing and storage techniques, in *Proc. Comput. Cardiol.*, pp. 191–194 (1978).
20. L. A. Lehmann and A. Macovski, Data compression of x-ray images by DPCM coding, *Proc. SPIE* **314,** 396–404 (1981).
21. P. C. Christenson, T. W. Ovitt, H. D. Fisher, M. M. Frost, S. Nudelman, and H. Roehrig, Intravenous angiography using digital video subtraction, *Am. J. Roentgenol.* **135,** 1145–1152 (1980).
22. R. L. Lillestrand, Techniques for change detection, *IEEE Trans. Comput.* **21,** 654–659 (1972).
23. P. Alexander, Array processors in medical imaging, *Computer (IEEE)* **16**(6), 17–31 (1983).
24. R. A. Kruger, A method for time domain filtering using computerized fluoroscopy, *Med. Phys.* **8,** 466–470 (1981).
25. R. G. Gould, M. J. Lipton, P. Mengers, and R. Dahlberg, Investigation of a video frame averaging digital subtraction fluoroscopic system, *Proc. SPIE* **314,** 184–190 (1981).
26. M. Pfeiler and P. Marhoff, Zur Technik der digitalen Röntgenbildvararbeitung, *Electromedica* **51**(1), 20–31 (1983).
27. K. H. Höhne, M. Böhm, W. Erbe, G. C. Nicolae, G. Pfeiffer, and B. Sonne, Computer angiography: A new tool for x-ray functional diagnosis, *Med. Progr. Technol.* **6,** 23–28 (1978).
28. J. H. Bürsch, H. J. Hahne, R. Brennecke, D. Grönemeyer, and P. H. Heintzen, Assessment of arterial blood flow measurements by digital angiography, *Radiology* **14,** 30–47 (1981).
29. R. Vogel, M. T. Le Free, E. Bates, W. O'Neill, R. Forster, P. Kirlin, D. Smith and B. Pill, Applications of digital techniques to selective coronary arteriography, *Am. Heart J.* **107,** 153–164, (1984).
30. R. Brennecke, H. J. Hahne, J. H. Bürsch, and P. H. Heintzen, Optimization of generalized subtraction operations for digital fluorography, in *Digital Imaging in Cardiovascular Radiology* (P. H. Heintzen and R. Brennecke, eds.), Thieme, Stuttgart, pp. 67–80 (1983).
31. R. A. Kruger, J. A. Nelson, F. J. Miller, P. Liu, W. Bateman, and S. Baron, Temporal filtering techniques for digital angiography, in *Digital Imaging in Cardiovascular Radiology* (P. H. Heintzen and R. Brennecke, eds.), Thieme, Stuttgart, pp. 80–88 (1983).
32. M. Böhm and K. H. Höhne, The processing and analysis of radiographic image sequences, in *Digital Image Processing in Medicine* (K. H. Höhne, ed.), Springer, Berlin, pp. 15–41 (1982).
33. J. A. Seibert, O. Nalcioglu, and W. W. Roeck, Deconvolution technique for the improvement of contrast of image intensifiers, *Proc. SPIE* **314,** 310–318 (1981).
34. C. G. Shaw, D. L. Ergun, P. D. Myerowitz, M. S. Van Lysel, C. A. Mistretta, W. C. Zarnstorff, and A. B. Crummy, A technique of scatter and glare correction for videodensitometric studies, *Radiology* **142,** 209–213 (1981).
35. R. Brennecke, T. K. Brown, J. H. Hürsch, and P. H. Heintzen, A digital system for roentgen-video image processing, in *Roentgen-Video Techniques* (P. H. Heintzen and J. H. Bürsch, eds.), Thieme, Stuttgart, pp. 150–157 (1978).
36. W. W. Peppler, M. S. Van Lysel, J. T. Dobbins, J. C. Lancaster, J. Hicks, B. H. Hasegawa, C. S. Lee, N. Sheikh, W. C. Zarnstorff, C. A. Mistretta, P. D. Myerowitz, and D. K. Swanson, Progress report on the University of Wisconsin digital video image processor (DVI II), in *Digital Imaging in Cardiovascular Radiology* (P. H. Heintzen and R. Brennecke, eds.), Thieme, Stuttgart, pp. 56–66 (1983).
37. B. K. Gilbert, M. T. Storma, C. E. James, L. W. Holbrock, E. S. Yang, K. C. Ballard,

and E. H. Wood, A real-time hardware system for digital processing of wide-band video images, *IEEE Trans. Comput.* **25,** 1089–1100 (1976).

38. P. Mengers, Real-time digital image processing for radiological imaging, in *Real-Time Radiologic Imaging* (D. A. Garrett and D. A. Bracher, eds.), Americal Society for Testing and Materials, Philadelphia, pp. 267–276 (1980).
39. R. H. McMann, S. Kreinik, J. K. Moore, A. Kaiser, and J. Rossi, A digital noise reducer for encoded NTSC-signal, *J. SMPTE* **87,** 129–133 (1978).
40. R. A. Kruger, C. A. Mistretta, J. Lancaster, T. L. Houk, M. Goodsitt, C. G. Shaw, S. J. Riederer, J. Hicks, J. Sackett, A. B. Crummy, and D. Fleming, A digital video image processor for real-time x-ray subtraction imaging, *Opt. Eng.* **17,** 652–657 (1978).
41. F. Buchmann and S. Balter, Roentgenbilder in digitaler Darstellung, *Computertomographie* **1,** 47–51 (1981).
42. M. S. Van Lysel, W. C. Zarnstorff, J. C. Lancaster, C. A. Mistretta, and J. T. Dobbins, III, Real-time digital video recording system, *Proc. SPIE* **314,** 389–395 (1981).
43. D. J. Hedberg, A real-time digital video disc recorder for medical imaging, *Proc. SPIE* **318,** 47–55 (1982).
44. C. Warren, A mix of standard and proprietary buses marks the latest microcomputer systems, *Electron. Des.* **17,** 114–120 (1981).
45. T. Wendler, D. Böhring, D. Meyer-Ebrecht, J. Schmidt, and H. Svensson, Modular multiprocessor picture computer architecture for distributed picture information systems, *Proc. SPIE* **318,** 125–132 (1982).
46. W. Ameling, Computer structures for digital imaging, in *Advances in Noninvasive Cardiology* (J. Meyer, P. Schweizer, and R. Erbel, eds.), Nijhoff, The Hague, pp. 1–12 (1983).
47. M. Onoe, K. Preston, and A. Rosenfeld (eds.), *Real-time Medical Image Processing,* Plenum Press, New York (1980).
48. G. C. Nicolae and K. H. Höhne, Multiprocessor system for real-time digital processing of video-image series, *Elektron. Rechenanlagen* **21,** 171–183 (1979).
49. R. Brennecke, H. J. Hahne, and P. H. Heintzen, A multiprocessor system for the acquisition and analysis of video image sequences, in *Erzeugung und Analyse von Bildern und Strukturen* (S. J. Pöppl and H. Platzer, eds.), Springer, Berlin, pp. 113–122 (1980).
50. P. C. Treleaven, D. R. Brownbridge, and R. P. Hopkins, Data-driven and demand-driven computer architecture, *ACM Comput. Surv.* **14,** 93–143 (1982).
51. P. H. Heintzen, R. Brennecke, and J. H. Bürsch, Present status of digital angiocardiography, in *Advances in Noninvasive Cardiology* (J. Meyer, P. Schweizer, and R. Erbel, eds.), Nijhoff, The Hague, pp. 51–66 (1983).
52. ACR-NEMA digital imaging and communication standard draft, November 19 (1984).
53. R. Brennecke, D. Jung, W. Clas, R. Erbel and J. Meyer, TALISMAN, An interpreter language for image sequence management, *Proc. Comp. Cardiol. 1985,* IEEE Computer Society, Long Beach, California (in press).

3

Temporal Integration Processing Techniques

S. J. Riederer

1. INTRODUCTION

The last several years have seen the introduction and rapid development of digital subtraction angiography (DSA), a technique which has been used to study the intracranial, extracranial, peripheral, and abdominal vessels, and the cardiac chambers. Although DSA techniques are by no means uniform, imaging rates for those first four anatomic regions are typically one image per second. Generally, a sequence of 10 to 15 short, intense exposures is made and only one or two of the resultant difference images are ultimately used. Consequently, there is significant patient exposure that does not result in a final diagnostic image. As an alternative, Kruger[1] and Gould *et al.*[2] have independently proposed imaging schemes that employ recursive digital temporal filtering in conjunction with low-intensity, continuous X-ray exposures. Such methods appear viable in some applications. As an alternative, Kruger and Liu[3] and Riederer *et al.*[4] have proposed the concept of "matched filtering" (MF) as a more efficient means of generating DSA images. Additionally, Enzmann *et al.*[5] and Maier *et al.*[6] have studied simple temporal integration techniques. This chapter is a review of the theory and applications of this technique.

S. J. Riederer • Department of Radiology, Duke University Medical Center, Durham, North Carolina 27710.

2. THEORY

2.1. The Conventional DSA Reference

The technical characteristics of MF have been discussed in the literature.[3,4,6,7] Moreover, it has been shown that MF is a special case of generalized temporal subtraction filters.[8] This section is a review of the major theoretical conclusions.

In many applications of DSA, the exposure rate used is one per second, and exposures are made over run times as long as 15–20 sec. This duration spans the arrival, peak opacification, and washout stages of the contrast material bolus. In many studies, only one or two difference images are ultimately used, such an image typically being formed from exposures made at peak and no opacification. As a reference we designate the 1 image/sec, simple (nonintegrated) difference technique as conventional DSA (CDSA).

CDSA is illustrated graphically in Figure 1. At the top of the figure is a hypothetical body portion containing an artery of interest and a mobile noniodinated object such as bowel gas or the larynx. Below this is shown a plot of DSA values across a line of the difference image made through the body portion at the time of peak iodine contrast. Depicted are signals from the iodinated artery (S), motion artifacts (A), and the peak-to-peak magnitude of the noise (σ). This plot will be used as a reference in discussing applications of integration techniques.

2.2. Temporal Integration

A common technique for reducing noise in many experimental studies is integration. Several or many measurements of the same phenomenon are added together, generally yielding a net signal which has increased linearly with the number of measurements and a net noise level which has increased in less than linear fashion. The ratio of the signal-to-noise

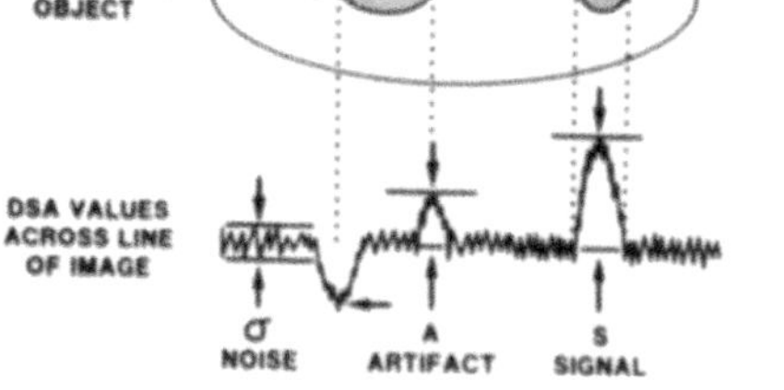

Figure 1. Schematic of a body section and signals across one line of a DSA difference image.

levels, or the net signal-to-noise ratio (SNR) improves. If the individual measurements used in the integration are uncorrelated, then the SNR improvement increases as $\sqrt{N}$, where N is the number of measurements.

This same technique can be applied to DSA, in which case a single measurement corresponds to one of the initial unsubtracted images. The signal is the slight additional decrease in video level caused by the presence of contrast material in the vessels of interest. By appropriately adding these unsubtracted images together, it is expected that some improvement in SNR can be obtained over the CDSA case.

Two characteristics of DSA that affect the actual implementation and performance of temporal integration should be pointed out. First, since DSA is a subtraction technique, it is still necessary that a subtraction be performed in the integration case. This is done by adding together several contrast-free images to form an integrated mask, doing likewise with contrast-filled images to form an integrated contrast image, and then taking their difference. By using averaging rather than integration for both mask and contrast images, it is possible to integrate different numbers of unsubtracted images for each. However, to benefit from integration the two numbers should not be markedly different.

The second point which should be discussed affects performance. Because the contrast bolus washes into and out of the vessels with a characteristic transit time of 5–6 sec, not all of the contrast-filled images will have the same contrast levels. This means that the SNR will not increase as rapidly as the $\sqrt{N}$ behavior alluded to earlier, because the signal is changing from image to image. Nevertheless, at a one per second imaging rate, integrating three or four images each for the integrated mask and contrast-filled images can yield a 50% improvement in SNR over the CDSA case. Clinical results have been reported by Enzmann *et al.*[5] The objective of MF is to perform more extensive temporal integration and obtain the best possible SNR improvement.

2.3. Matched Filtering

Consider a signal $b(t)$ of known shape contained within a noisy waveform. Through temporal filtering methods, it is desired to improve the detectability of $b(t)$. It is well known in signal processing that the filter which maximizes the SNR of the result has a shape proportional to, i.e., which "matches," that of the known signal. In intravenous angiographic imaging, one is dealing with bolus signals having typical widths of 5–10 sec. Furthermore, in the absence of appreciable motion artifacts, DSA is principally limited by the ratio of iodine signal to background noise level. Thus, DSA appears well suited to MF techniques.

The ability of DSA to visualize small contrast levels depends not only on having an adequate SNR in the difference image, but also in the subtraction step itself. In subtracting a precontrast mask image from a sequence of contrast-filled images, one is eliminating all static signals. MF techniques applied to DSA must not only yield a result with adequate SNR, but also have a temporal dc frequency response of zero.

Let $\{I_i\}$ be a set of unsubtracted video images and $\{h_i\}$ the temporal domain matched filter applied to the images. The filtered result is then expressed as

$$S_{\mathrm{MF}} = \sum_{i=1}^{N} h_i I_i \tag{1}$$

The dc response of the matched filter is zero if its discrete Fourier transform is zero at zero frequency:

$$\sum_{i=1}^{N} h_i = 0 \tag{2}$$

Rather than unsubtracted images, suppose that difference images are subjected to the filtering, each difference image the result of subtracting a live image from a common mask. Since all static image components are already canceled prior to filtering, it may seem that the requirement in Eq. (2) is no longer necessary. However, a derivation for the variance σ^2_{MF} in the MF result for this case yields

$$\sigma^2_{\mathrm{MF}} = \sigma^2 \sum_{i=1}^{N} h_i^2 + \sigma_{\mathrm{M}}{}^2 \left[\sum_{i=1}^{N} h_i \right]^2 \tag{3}$$

where σ_{M} and σ are the rms noise levels in the mask and live images, respectively. The first term is recognized as the noise level generally associated with linear filtering, while the second term is specifically from the mask image. In the set of difference images, the noise of the mask, although random, forms a fixed pattern present in all difference images. To eliminate this pattern, the filter dc response must be zero and Eq. (2) must still apply.

Because of variation of bolus dynamics from patient to patient, any bolus signal assumed prior to a procedure will in general not match that actually observed. In practice, this can be circumvented by first storing either the sequence of unsubtracted video images or simple difference images. After completion of the procedure, the bolus curve can then be estimated from either sequence.

Because the bolus signal is always nonnegative, and Eq. (2) implies that some filter values must be negative, one never has a true match of signal with filter. As an approximation, once the bolus is estimated from the image sequence, the filter is formed by subtracting the mean bolus value from all bolus points so that Eq. (2) is then satisfied.

Figure 2 illustrates the MF technique. Figure 2a is a plot of the iodinated contrast bolus observed in a vessel as a function of time fol-

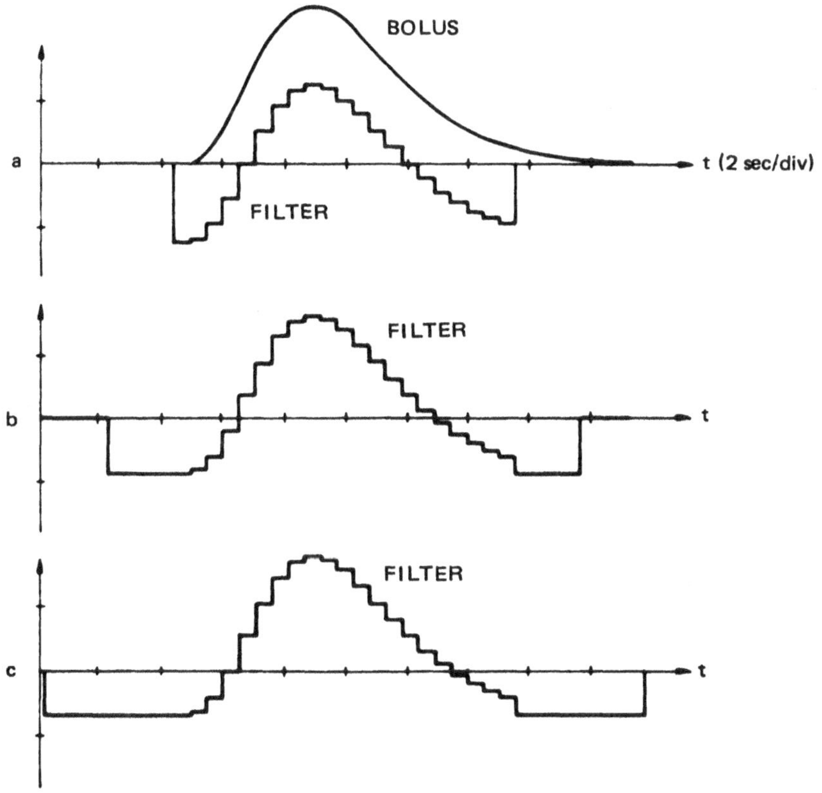

Figure 2. Demonstration of improved match of filter with bolus with increased exposure time. In (a), the exposure is limited to the time of the bolus. Images at onset and washout of the bolus must be negatively weighted so that the filter integral is zero. This negative weighting, however, reduces output signal size. As exposure duration is increased in (b) and (c), more of the bolus is positively weighted while the zero filter integral condition is maintained.

lowing intravenous injection. This signal would be observed in a set of simple difference images. Suppose that during this time, exposures were made at a 2/sec rate; the matched filter that would be derived from this data set is also shown. Images at the leading and trailing edges of the bolus curve would be weighted negatively, while those at or near peak contrast would have the largest positive weighting. The sum of all filter values would be zero, in accordance with Eq. (2). As shown in Figure 2b, the imaging duration t is extended and total detected exposure increases proportionately. The SNR_{MF} increases not only because of the increase in total exposure, but also since more of the bolus is positively weighted. As duration is increased even farther, as in Figure 2c, the added exposure becomes less effective in increasing SNR_{MF}. If a given total exposure is to be administered continuously within some time interval t for MF, then for the bolus of Figure 2a the highest SNR is obtained for $t \approx 14$ sec.

Analysis by Riederer *et al.*[4] has compared MF with CDSA and with a bound on DSA performance. The bound is the case in which all exposure administered in a DSA run is equally apportioned between a single con-

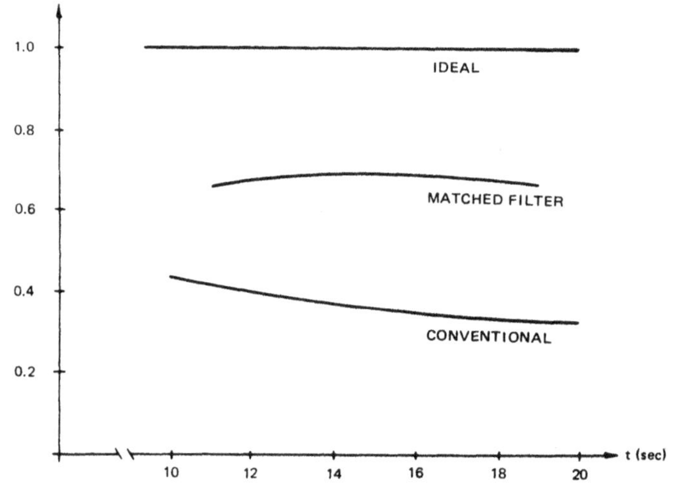

Figure 3. Relative signal-to-noise ratios (SNRs) of the ideal case, matched filtering, and conventional DSA as a function of DSA exposure duration. For exposures over times of 13 to 17 sec, matched filtering typically provides an SNR twofold higher than conventional DSA. Over the same times, the SNR of the optimum is only 1.4-fold higher than matched filtering. A typical recursive filter method would have a relative SNR $7.30/\sqrt{30}$ or 1.3-fold higher than conventional DSA. For example, at 15 sec the relative SNR would be 0.48.

trast-free image and a second image at peak contrast. Figure 3 compares the relative SNRs of these three cases plotted as a function of the duration of the DSA run. Note the marked improvement provided by MF over CDSA. For example, at $t = 15$ sec, the relative SNRs are 0.70 and 0.36. This means that for a given total exposure administered over 15 sec, the MF method provides an SNR that is a factor of 2 better than that of CDSA. Equivalently, CDSA would require an exposure rate that is a factor of $(0.70/0.36)^2$ or about 4-fold higher to attain an SNR comparable to that of MF. For this same 15-sec duration, results from a typical recursive filter would have a relative SNR of $7.30/\sqrt{30}$, or about 1.3 × higher than conventional DSA. It too would provide some exposure reduction versus CDSA for equal SNR.

One is reassured by the performance of MF with respect to the ideal case. Although the latter is woefully impractical, MF requires only a factor of $(1/0.7)^2$ or 2-fold higher exposure to achieve equal SNR performance. Thus, we conclude that MF not only provides considerable improvement in DSA compared to conventional methods, but also is close to an absolute bound on performance.

3. IMPLEMENTATION

Many of the components necessary for performing temporal integration or MF are already found in most real-time DSA systems. The process is implemented by first performing a DSA run and acquiring a sequence of unsubtracted or simple difference images. Next, upon viewing a simple difference image, the observer selects with a cursor some region-of-interest (ROI) such as a portion of a vessel of interest. The DSA processor next calculates the signal level in the region for each image of the original sequence, yielding a set of values analogous to the bolus curve of Figure 2a. The actual matched filter is formed by subtracting the average from all of the measured values. Finally, the matched filtered image is generated by taking the linear combination of the original images, the coefficient or "weight" being equal to the filter value. This step can be performed by recalling each original image from storage, and using appropriate high-speed digital circuitry to perform the weighting and add all weighted images together. Processor requirements which are potentially over and above those required for routine DSA are the ability to perform the weighting or multiplication and precision adequate for encoding a linear combination of up to 50 images.

Because motion artifacts are one of the principal limitations of CDSA, it is natural to question how the extensive temporal integration of MF is

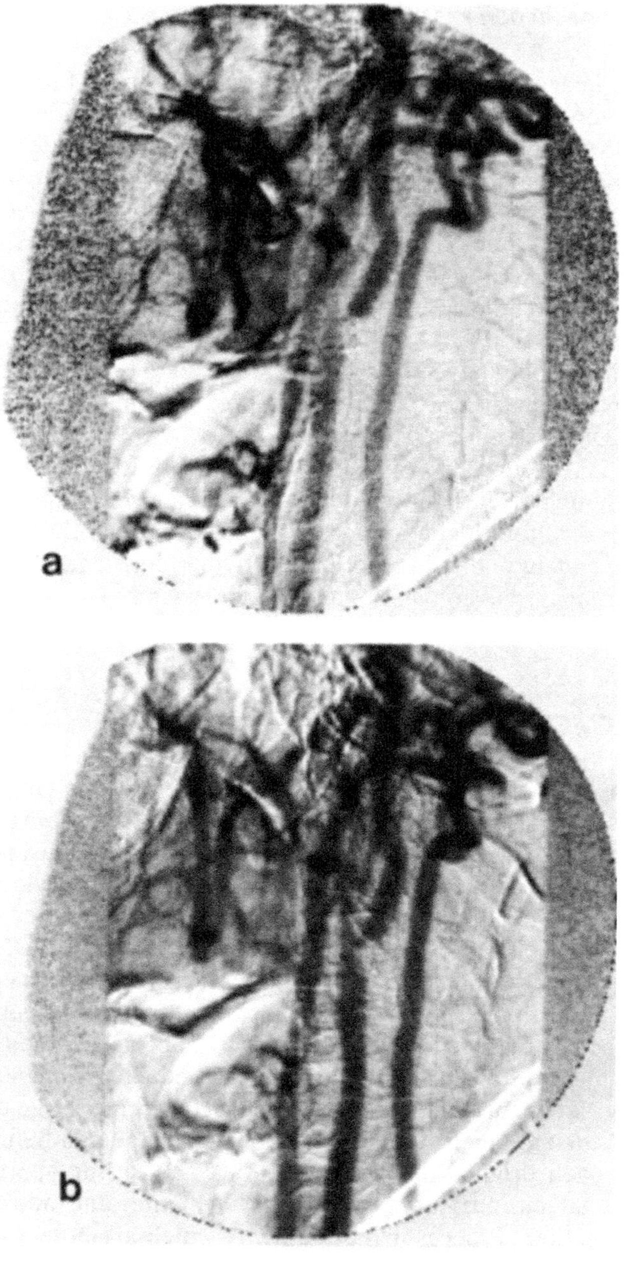

Figure 4. (a) DSA image of left carotid artery bifurcation done in 60° LAO projection using conventional 1/sec technique. (b) Matched filtered image formed using images 4 through 30 of the available 38 images. Misregistration artifact is exactly at the area of interest. (c) Matched filtered image generated using images 23 through 38 of the same image sequence used for (b). Note virtual elimination of motion artifact at area of interest and clear delineation of tight stenosis. (d) Schematic representation of bolus curve and filters for (b) and (c).

c

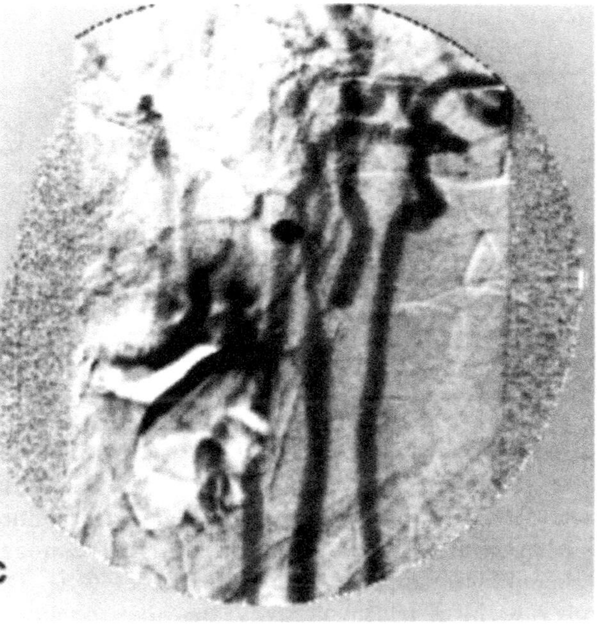

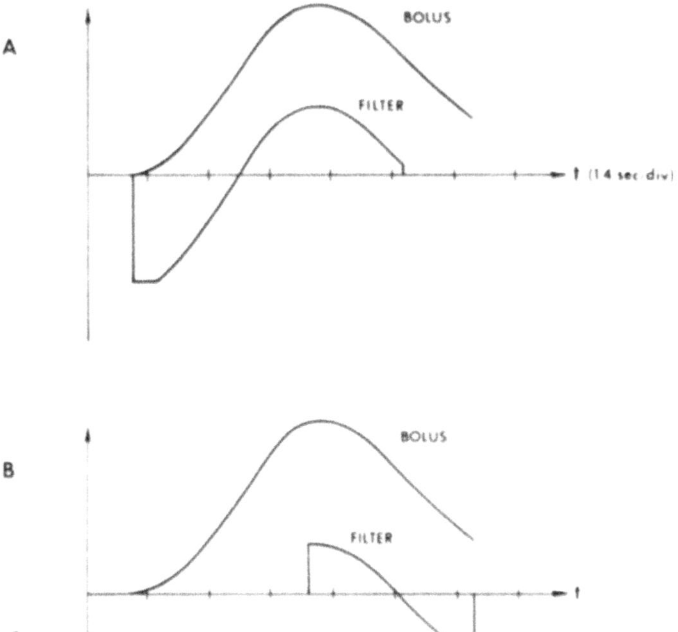

d

(A) ''Bolus'' is a smooth curve fit to the DSA values in a region of interest in the left carotid bifurcation. Images were acquired throughout the 8- to 9-sec duration of the bolus at a 4/sec rate. ''Filter'' is a plot of the matched filter. (B) ''Bolus'' as in (A). The initial nonzero filter value is at a time just after appearance of motion artifact. Positive and negative portions of filter are equal, causing all static image features, such as stationary motion artifacts, to be eliminated in final result.

even possible without artifact. Motion sensitivity of the MF method can be reduced in the following fashion. If only half of the available raw difference images are usable because the remainder have been corrupted by motion artifact, the MF algorithm can be tailored to use only these good images.

An example of using only a subset of the original images for MF is shown in Figure 4. The result of a CDSA (one exposure per second) run is shown in Figure 4a. A tight stenosis at the origin of the left internal carotid artery is demonstrated in this LAO projection. An MF run was next done at four images per second and an exposure per image that was one fourth that of the first run, so that total exposure for the two runs was identical. Using images 4 through 30 of the 38 available images yielded the MF result shown in Figure 4b. Unfortunately, a motion artifact obscures the area of interest. The gamma variate curve modeling the bolus and the MF for this image are shown in Figure 4d(A). Playback of the sequence of raw difference images revealed that the motion artifact occurred near peak opacification and remained relatively stationary thereafter. An MF was formed that excluded the images that preceded the motion, and it is shown in Figure 4d(B). In the resultant MF image of Figure 4c, the artifact has been virtually eliminated. Moreover, the SNR in the area of interest appears superior to that in the CDSA result of Figure 4a. Although there was no exposure advantage over DSA, this case provides an example of selective MF. It should be noted that the motion artifact was present in each of the raw difference images used for filtering. Similar to remasking in DSA, an artifact that is constant in a set of images can be removed in the MF result.

To perform MF on such edited DSA image sequences, it is desirable to recall arbitrary original images very rapidly. Additionally, motion detection algorithms would be useful in helping determine which images to discard.

4. APPLICATIONS

4.1. SNR Improvement

Perhaps the most direct application of MF is SNR enhancement, shown schematically in Figure 5, in which case exposure rates and contrast material amounts identical to CDSA are used. At the top of the figure is a plot of the DSA values across one line of an original difference image, as acquired (ACQ) during the actual passage of contrast material through the patient. Note that it is identical to the CDSA case at the bottom of

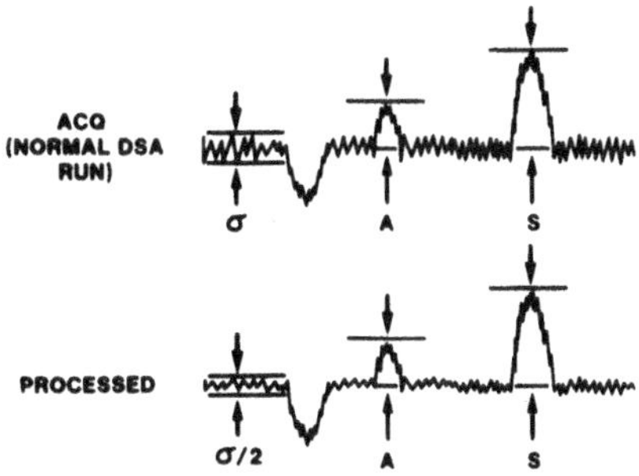

Figure 5. Schematic of SNR improvement.

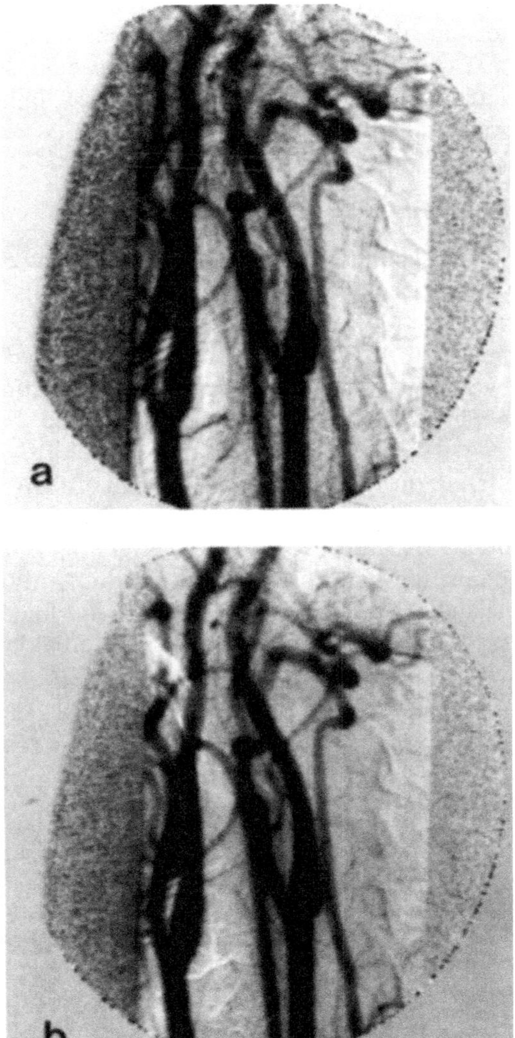

Figure 6. Original CDSA image (a) and matched filtered result (b) demonstrating SNR improvement.

Figure 1. The same line, but now of the matched filter processed image, is also portrayed in Figure 5 (PROCESSED). Note that the noise magnitude is now 50% lower than it was during acquisition and, accordingly, the SNR is twice as high. Potentially smaller contrasts can be perceived in the result compared with CDSA. However, since presumably the SNR of CDSA was adequate in the first place, the twofold improvement of SNR provided by MF in this application may be principally cosmetic.

An example of SNR enhancement is given in Figure 6, an intravenous DSA study of the extracranial carotid arteries. A simple difference image (the CDSA result) is shown in Figure 6a, demonstrating excellent opacification of both carotid bifurcations. All 15 images from this run were next matched filtered, yielding Figure 6b. Note that the noise level, particularly in the upper portions of the image, is noticeably reduced as predicted from Figure 5.

4.2. X-Ray Exposure Reduction

A second application of MF, X-ray exposure reduction, is shown schematically in Figure 7. The idea of this application is as follows. Because the MF processing reduces noise by a factor of 2, during acquisition one, in principle, need only obtain images with a twofold larger noise level than normal. This situation in acquisition is shown in Figure 7 (ACQ), in which case the noise magnitude is now 2σ while the iodine signal (S) and artifact (A) are the same as before. After the image sequence is filtered, the resultant schematic plot of DSA values is shown on the bottom of Figure 7. Note that the iodine SNR S/σ is identical to that in Figure 1 for CDSA. Because a twofold larger noise level can be accommodated in acquisition with this application, the X-ray exposure can be reduced by the square of the noise increase, or fourfold. This approximate factor was corroborated experimentally in our initial investigations of MF in carotid and cerebral studies.[9]

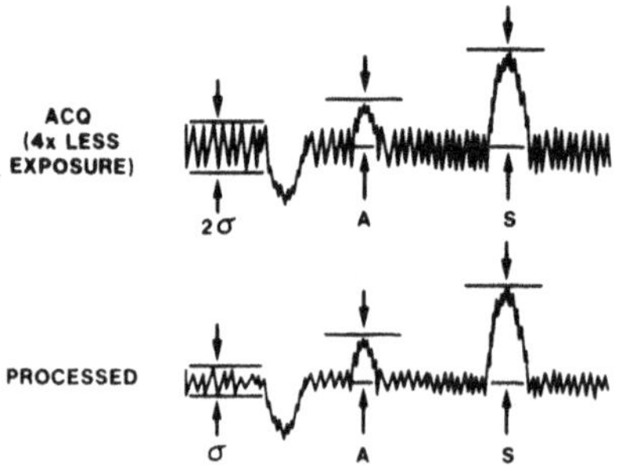

Figure 7. Schematic of X-ray exposure reduction afforded by matched filtering.

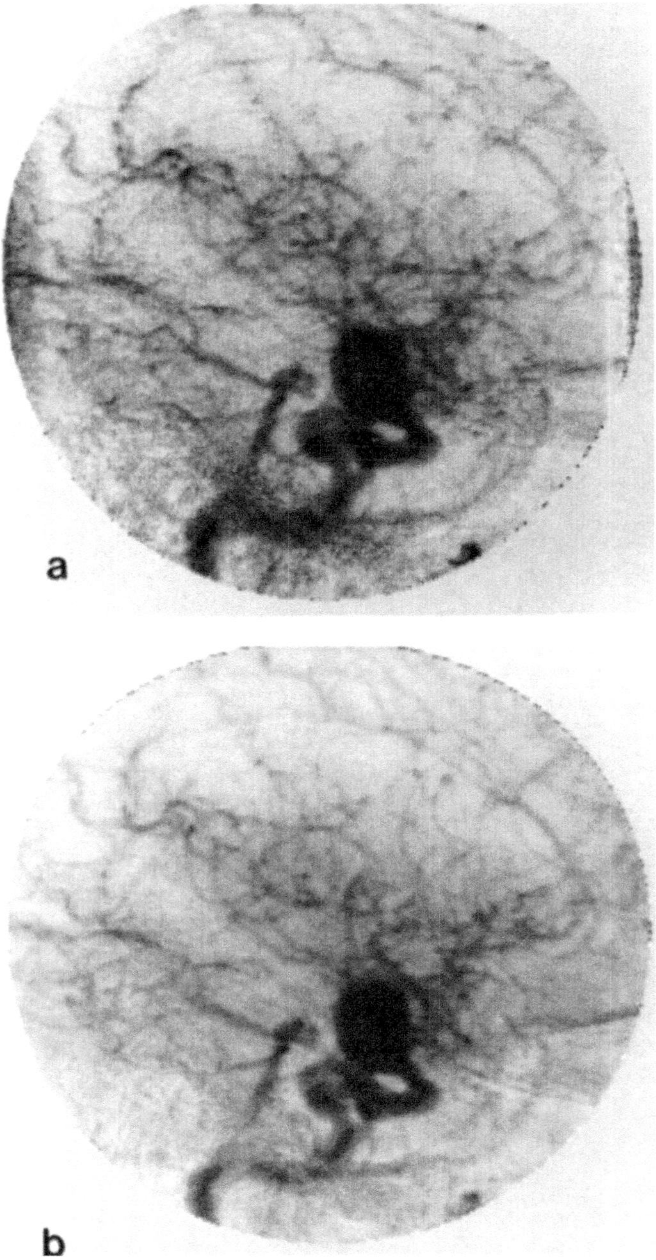

Figure 8. (a) CDSA image of intracranial aneurysm. (b) Matched filtered result of same anatomy as (a) but with a sixfold reduced exposure.

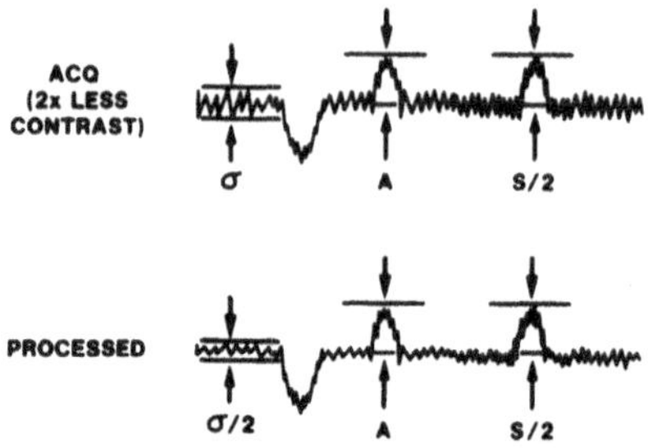

Figure 9. Schematic of contrast dose reduction.

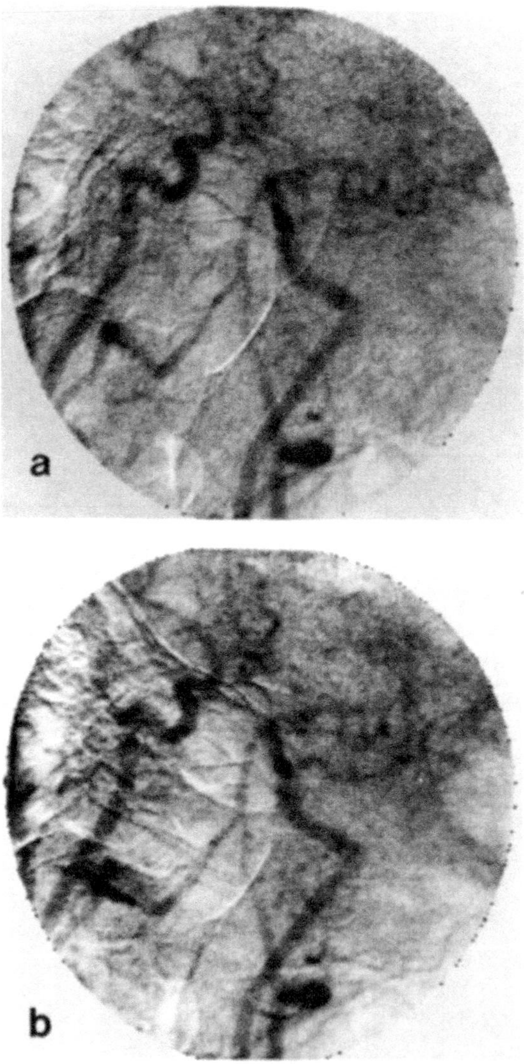

Figure 10. (a) CDSA image of vertebral–basilar system using 40 ml of Hypaque-76. (b) Matched filtered result of same anatomy as (a) with identical exposure and 20 ml.

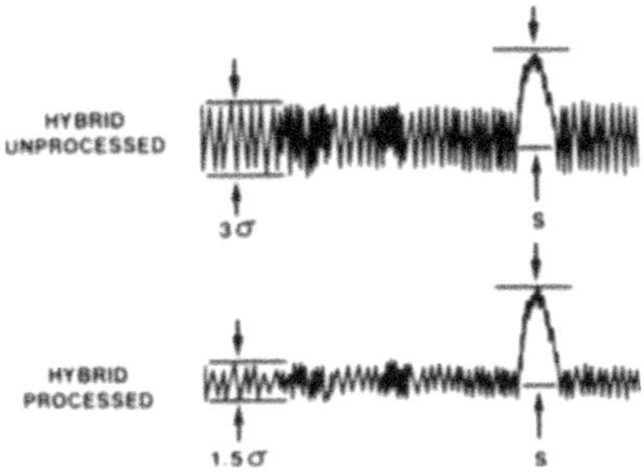

Figure 11. Schematic of matched filtering applied to hybrid subtraction.

The study in Figure 8 demonstrates X-ray exposure reduction. Figure 8a is the CDSA result of an intravenous study of a patient with an intracranial aneurysm. The study was next repeated with an identical injection technique but with a sixfold reduced X-ray exposure per second. The result of matched filtering the entire 14-sec low-exposure run is shown in Figure 8b, demonstrating an SNR that is substantially equivalent to that of the CDSA result.

4.3. Contrast Dose Reduction

Contrast material dose reduction, a third application of MF, is portrayed in Figure 9. The intent in this case is to reduce the contrast material dose by a factor of 2 while using the same X-ray exposure rates as CDSA. Thus, in the acquisition (ACQ) schematic of Figure 9, the signal (S/2) is one-half of that in Figure 1, while the noise magnitude (σ) is the same for both. After processing (bottom of Figure 9), the magnitude is reduced to $\sigma/2$. With the 50% reduction in both signal and noise, the final SNR is given by $(S/2)/(\sigma/2) = S/\sigma$, which is identical to the SNR of CDSA in Figure 1.

In discussing Figures 1, 5, 7, and 9, little mention was made of the brightness due to the motion artifact (A). Assuming the same degree of motion for each, a valid measure of the relative sensitivity to motion of the three MF applications is the ratio of artifact magnitude to iodine signal. For SNR enhancement and X-ray exposure reduction (Figures 5 and 7), this is seen to be A/S. On the other hand, for the contrast material dose reduction case (Figure 9), the ratio is A/(S/2) or 2A/S, which is twice as high. We thus conclude that motion sensitivity is twice as severe for the contrast material dose reduction case.

Contrast dose reduction in DSA is illustrated in Figure 10. The CDSA result, formed with a 40-ml intravenous injection of Hypaque-76 (370 mg

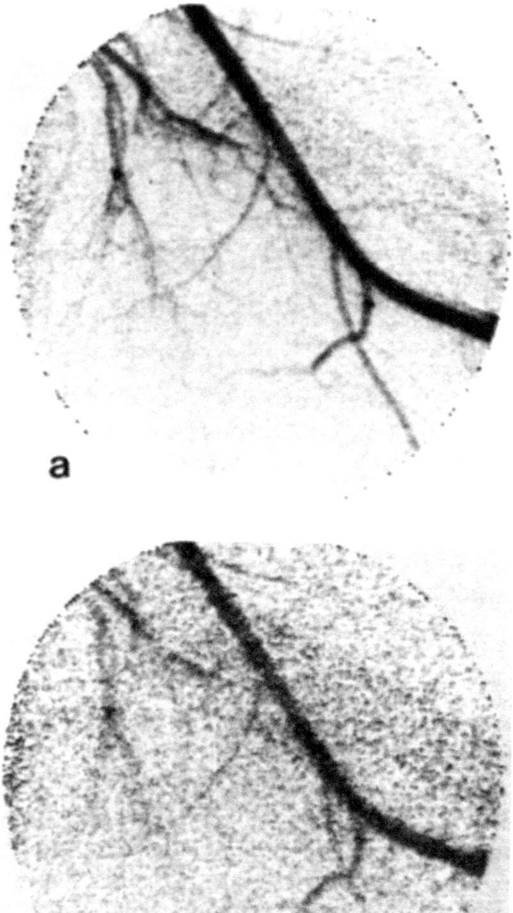

Figure 12. (a) CDSA image of the femoral arteries of a dog. (b) Unprocessed hybrid subtraction image illustrating significant degradation in SNR. (c) Matched filtered hybrid subtraction image illustrating SNR recovery in comparison to (b).

I/ml), depicts the vertebral–basilar system. The study was then repeated with an identical X-ray technique but with only a 20-ml contrast injection. The resultant image sequence was matched filtered, yielding the image in Figure 10b, substantially equivalent to that in Figure 10a. Our clinical

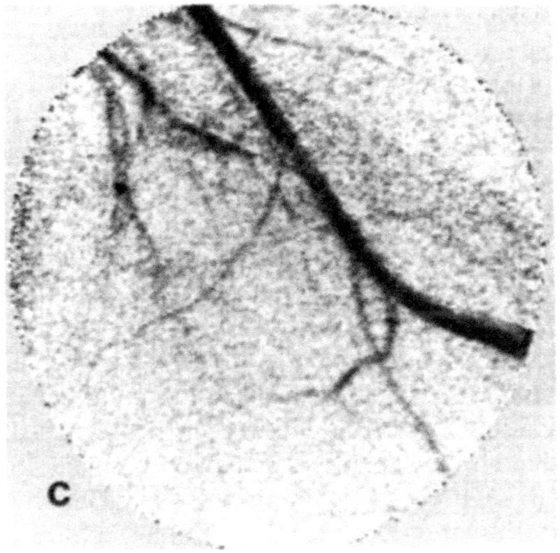

Figure 12. (*continued*)

studies of the half-contrast dose technique have been discussed in the literature.[10]

4.4. Hybrid Subtraction SNR Recovery

The combination of energy and temporal or "hybrid" subtraction is a technique proposed by Brody[11] to address the problem of soft tissue motion in temporal subtraction DSA. Unfortunately, the technique suffers from SNR degradation (2.5–3.0 ×) in comparison to temporal subtraction.[12] However, integration techniques such as MF can be used to recover some of this lost SNR.[13]

For diagrammatic purposes, one can envision hybrid subtraction as a sequence of images simultaneously formed using low- and high-kVp beams. Each high-kVp exposure is made very shortly (30–50 msec) before or after a corresponding low-kVp exposure. A plot of DSA values for one line of the image similar to that in Figure 1 would result for the high-kVp exposure except that the iodine signal (S) would be reduced substantially and the artifacts (A) would be reduced somewhat because of the higher beam energy. These low- and high-kVp difference images are then subtracted from each other in such a fashion as to eliminate the motion artifact (A). The plot of DSA values for this unprocessed hybrid image is shown in Figure 11. Note that the artifact has been eliminated, but that the background noise has increased, yielding an SNR reduced threefold relative to CDSA. Depending on the iodine levels, this SNR reduction may be unacceptable.

All pairs of high- and low-kVp exposures can be combined, yielding a sequence of unprocessed hybrid images that spans bolus arrival and washout. When this sequence is matched filtered or extensively integrated, the resultant processed hybrid image has a noise level reduced twofold compared with its unprocessed counterpart. This is shown schematically in the DSA plot, corresponding to hybrid (processed), of Figure 11. Thus, the temporal integration step has restored two-thirds of the SNR that was lost in going from temporal to hybrid subtraction. With an additional increase (~ twofold) in patient X-ray exposure, approximate parity with the temporal subtraction results can be attained.

This application of MF is illustrated in Figure 12. Although there are no motion artifacts in the temporal subtraction image, the SNR recovery is demonstrated. Figure 12a is the intravenous CDSA result of a study of the femoral arteries of a dog. Figure 12b is an unintegrated hybrid subtraction image, formed from Figure 12a and an analogous high-kVp difference image. The noise level is significantly higher as discussed above, but any tissue motion artifacts which would have been present in Figure 12a would be substantially eliminated. Finally, Figure 12c is the result of matched filtering the hybrid subtraction images, demonstrating substantial SNR recovery back toward the SNR of the CDSA case in Figure 12a.

5. DISCUSSION

The underlying theory of MF in DSA has been reviewed and examples of several applications have been given. As stated at the outset, MF represents a trade-off between SNR improvement and reduced temporal resolution. Perhaps the major limit of the MF technique is motion during the integration interval. Motion artifacts may come from two distinct sources: artifacts from moving noniodinated structures, and distortion from vessel motion during the integration interval. If extreme enough, motion will eventually corrupt all temporal subtraction schemes. However, these two artifacts may be addressed respectively by "editing" an image sequence prior to input to the matched filter[9] as illustrated in Figure 4 and by using EKG gating techniques. Instrumentation required to perform MF is not appreciably different from that for routine DSA. Enough storage and bits of precision are required to form linear combinations of a sequence of images.

ACKNOWLEDGMENT. The author wishes to thank Debra Kaufman for typing the manuscript.

REFERENCES

1. R. A. Kruger, A method for time domain filtering using computerized fluoroscopy, *Med. Phys.* **8,** 466 (1981).
2. R. G. Gould, M. J. Lipton, P. Mengers, and R. Dahlberg, Investigations of a video frame averaging digital subtraction fluoroscopic system, *Proc. SPIE* **314,** 189–191 (1981).
3. R. A. Kruger and P. Y. Liu, Digital angiography using a matched filter, *IEEE Trans. Med. Imag.* **1,** 16 (1982).
4. S. J. Riederer, A. L. Hall, J. K. Maier, N. J. Pelc, and D. R. Enzmann, The technical characteristics of matched filtering in digital subtraction angiography, *Med. Phys.* **10,** 209 (1983).
5. D. R. Enzmann, W. T. Djang, S. J. Riederer, W. F. Collins, A. L. Hall, G. S. Keyes, and W. R. Brody, Evaluation of peripheral vascular disease using digital subtraction angiography, *Radiology* **147,** 393–398 (1983).
6. J. K. Maier, S. J. Riederer, D. R. Enzmann, and W. R. Brody, Some advanced concepts of matched filtering in intravenous digital subtraction angiography, *Invest. Radiol.* **19,** 424–431 (1984).
7. D. L. Parker, P. D. Clayton, and L. R. Tarbox, Optimal dose utilization with variable x-ray intensity in digital radiography, *Proc. SPIE* **419,** 102–109 (1983).
8. R. A. Kruger and S. J. Riederer, *Basic Concepts of Digital Subtraction Angiography,* Hall, Boston (1984).
9. S. J. Riederer, D. R. Enzmann, A. L. Hall, N. J. Pelc, and W. T. Djang, The application of matched filtering to x-ray exposure reduction in digital subtraction angiography: Clinical results, *Radiology* **146,** 349–354 (1983).
10. S. J. Riederer, D. R. Enzmann, W. R. Brody, and A. L. Hall, The application of matched filtering to contrast material dose reduction in digital subtraction angiography, *Radiology* **147,** 853–858 (1983).
11. W. R. Brody, Hybrid subtraction for improved arteriography, *Radiology* **141,** 828–830 (1981).
12. G. S. Keyes, S. J. Riederer, B. F. Belanger, and W. R. Brody, Hybrid subtraction in digital fluorography, *Proc. SPIE* **347,** 34–41 (1982).
13. S. J. Riederer, W. R. Brody, D. R. Enzmann, A. L. Hall, and J. K. Maier, Work in progress: The application of temporal filtering techniques to hybrid subtraction in digital subtraction angiography, *Radiology* **147,** 859–862 (1983).

4

Noise Analysis in Digital Radiography

B. A. Arnold

1. INTRODUCTION

Noise plays a dominant role in the detection of low-contrast objects in any imaging system. Until the advent of the computed tomography (CT) scanner and more recently digital radiography, noise in an imaging system (typically a screen–film combination) was considered to have two major components: quantum mottle, the statistical fluctuations in the number of detected photons, and structure noise, due to phosphor screen inhomogeneities and film grain. The development of electronic X-ray detectors and in particular digital video subtraction systems has added much complexity to noise analysis. Due to the many new noise sources present, noise analysis now requires new evaluation criteria and techniques.

The rapid development of digital subtraction angiography in the past few years has required the use of new and improved imaging components.[1–12] These improvements have been focused heavily upon noise reduction particularly in the TV system.[13] The signal-to-noise ratio (SNR) of the TV camera required significant improvements before digital video angiography with intravenous (IV) injections could be implemented. The noise in the imaging chain which ends up in the output of the system, the final image, deteriorates the quality of the final image in both a subjective and a quantitative sense. This noise may lead to an undetected lesion in a visual interpretation reading or may create errors in quantitative results, such as percent stenosis or regional blood flow.

B. A. Arnold • Image Analysis, Inc., Irvine, California 92714, and Department of Radiological Sciences, UCLA Medical School, Los Angeles, California 90024.

The importance of the SNR with regard to image quality in IV subtraction angiography can be seen from the images shown in Figure 1. This set of images, in addition to showing the subjective importance of SNR, traces some of the historical developments obtained in our laboratory beginning in 1979. The subjective relationship between image quality and SNR in these images is very apparent although no attempt was made to standardize patients or exposure techniques. TV camera electronic noise was the dominant factor limiting the SNR in the first three images and a contributing factor in the last image in Figure 1.

In this chapter, the noise sources and characteristics of digital imaging systems will be discussed. Although the focus will be on digital video subtraction angiography systems, the principles and concepts will apply in most cases to all digital radiography systems. The sources of noise present in a digital radiography system will first be identified and discussed. A theoretical analysis of the effects of noise on signal detection will next be presented. Knowledge of the sources of noise and the parameters which control their magnitude will add physical meaning to the theoretical analysis. Finally, experimental results on a state-of-the-art digital video radiography system will be presented.

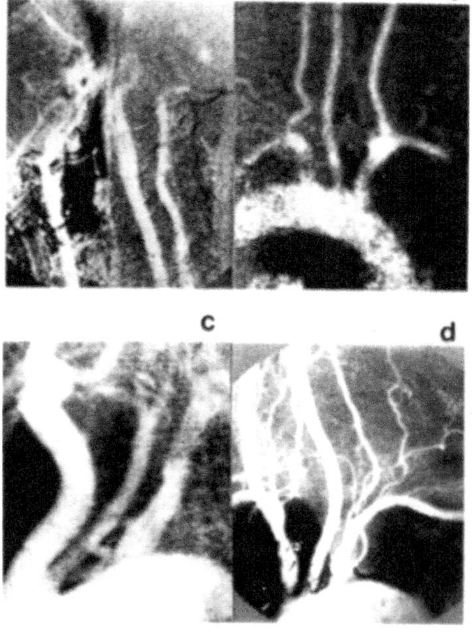

Figure 1. Clinical IV subtraction angiograms obtained during the early developmental stages of digital radiography. The signal-to-noise ratios of the imaging systems used during these early clinical trials were (a) SNR = 20:1, (b) SNR = 150:1, (c) SNR = 320:1, and (d) SNR = 750:1.

2. SOURCES OF NOISE IN DIGITAL SYSTEMS

The major sources of noise generation in a digital radiography system are listed in Table 1. Several of these noise sources have been recognized and evaluated to some degree by Kruger *et al.*,[1] Riederer *et al.*,[2] Frost *et al.*,[3] Roehrig *et al.*,[4] Scheibe and Thomas,[6] and Arnold *et al.*[7,8]

X-ray quantum noise is a fundamental noise source which is determined by the radiation dose that is delivered to the patient as well as the beam energy and absorption properties of the imaging detector and patient. In an ideally designed imaging system, all noise sources except quantum noise would be absent. Further, if individual photons could be counted in this idealized case, the noise properties of X-ray imaging systems could easily be characterized by Poisson statistics. In Section 4 (Theoretical Analysis), sample calculations will be carried out using these simplifying assumptions. However, in this section we deal with the real and more complex case where the energy deposited by absorbed photons is integrated, converted to an electronic signal, and then digitized. This is characteristic of all X-ray imaging systems in use at present including CT, line-scanned radiography, and digital video systems.

TV camera electronic noise is contributed primarily from preamplifier and beam shot noise in the camera tube. It is determined by the video bandwidth and by the design of the TV camera tube and camera electronics.

It will be useful to consider the generation of signal and noise in an X-ray image intensifier TV system. Light generated in the image intensifier tube is collected by the output optics and focused onto the photocathode of the TV camera. Even though the photoconductive layer is continuous, it can be considered as a two-dimensional array of target elements or pixels. The effective pixel size is proportional to the square of the effective beam diameter and the effective layer thickness. The electron beam of

TABLE 1. Sources of Noise in a Digital Video Radiography System

X-ray quantum noise
TV camera electronic noise
Quantization noise
Time jitter noise
Structure noise in image intensifier
Electronic noise of video recorders
Structure noise—time jitter interplay
Patient absorption inhomogeneities—patient motion
Scattered radiation

the TV camera scans this array of pixels sequentially, and the signal current measured at the TV camera output is determined by the number of electrons deposited onto a pixel during the dwell or exposure time on that particular pixel and by the electronic charge. Both the signal and TV noise sources are dependent on the readout time, i.e., the camera bandwidth. The bandwidth is given by

$$\text{Bandwidth} = N/2K(t_f - t_b) \tag{1}$$

where t_f is the readout time, t_b is the blanking time, N is the number of pixels, and K is the Kell factor. Typical readout times are 33 msec at bandwidths of 5–10 MHz. Equation (1) indicates that a TV camera can be optimized for a particular imaging situation depending upon the requirements for spatial and time resolutions and electronic SNR.

An additional source of noise unique to digital image systems is quantization noise. When digitized in the analog-to-digital converter, a function or signal is assigned one of several discrete numeric values dictated by the number of bits in the digitization word. All sample amplitudes occurring within any specific range are assigned a single value, i.e., one of two levels Q, either the level higher or lower than the sample value. Some of these samples may differ by as much as a full Q level. Figure 2 is a sketch of the digitization process. Quantization error is random in nature, and, under conditions usually met in practice,[14] is characterized by a rectangular probability distribution function. The magnitude of quanti-

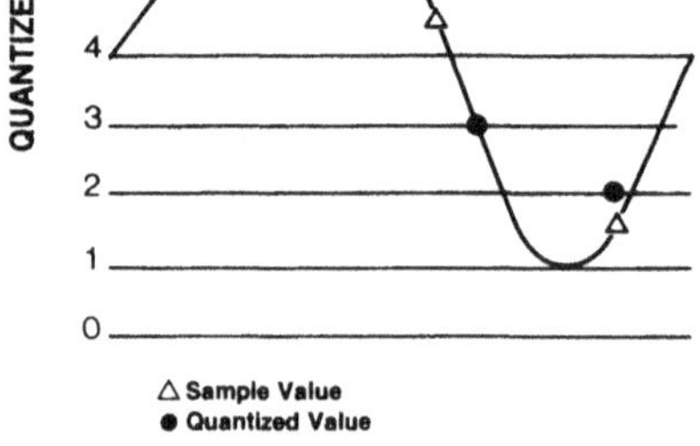

Figure 2. Generation of quantization error noise in digital systems. Sample (true) values are assigned discrete quantized values, which may differ by up to one Q level. This process is random, producing a noise source unique to digital systems.[8]

TABLE 2. Signal-to-Noise Ratios for Quantization Noise[8]

No. of bits, n	No. of levels, 2^n	SNR[a] (dB)
6	64	47
7	128	53
8	256	59
9	512	65
10	1024	71

[a] Peak-to-peak signal/rms noise.

zation error noise can therefore be calculated in a straightforward manner, and is given by

$$E_q = Q/\sqrt{12} \tag{2}$$

where E_q is the rms value of the quantization error and Q is the difference between adjacent quantization levels. For example, for an 8-bit computer which has 256 levels, the SNR for quantization noise is equal to 59 dB. This SNR represents peak-to-peak signal divided by rms noise for a full-range input signal. Table 2 shows similar calculations for other digitization accuracies.

Time jitter noise is due to time instabilities in components of the imaging chain, such as jitter in the position of the electron beam of the TV camera or time jitter in the analog-to-digital converter. Time jitter noise has increased in importance as imaging systems have improved in terms of other noise properties.

Image intensifier structure noise is due to inhomogeneities in X-ray absorption in the input phosphor and light emission of the output phosphor of the image intensifier. These irregularities produce localized variations in the signal intensity across the image plane in unsubtracted images. Image intensifier structure noise can be readily observed by recording a single video line with an oscilloscope while the image intensifier–TV camera is under X-ray exposure. Figure 3 shows such a measurement. The top trace was recorded with an input exposure to the image intensifier of 0.1 mR/TV frame, in which case quantum mottle is the dominant noise source. When the X-ray exposure is increased to 5.4 mR/frame, quantum noise becomes less significant. The structural inhomogeneities of the image intensifier can now be observed. The intensity variations shown in trace B are due to image intensifier structure noise and are fixed in time,

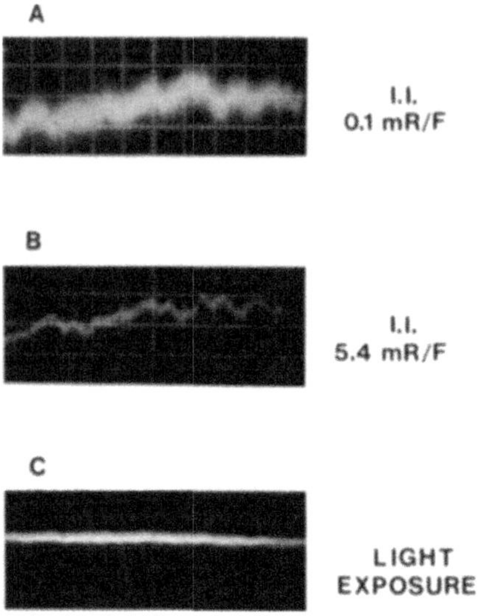

Figure 3. Oscilloscope traces showing image intensifier structure noise. (A) Low-input X-ray exposure shows high quantum noise level. (B) High-input exposure shows image intensifier structure noise pattern. (C) TV camera viewing homogeneous light source confirms that (B) is due to image intensifier.[8]

i.e., fixed pattern noise. The last trace, recorded with the TV camera viewing a while light source, confirms that this noise source is due to the image intensifier and not the TV camera. The fuzziness or blur seen in traces B and C, approximately 2 mV peak-to-peak, can be attributed to TV camera electronic noise.

Video recording devices have several electronic noise sources as discussed by Riederer *et al.*[2] The magnitude of video recorder noise is dictated primarily by the recording bandwidth and the engineering design of the electronics.

There exists a new set of noise sources unique to subtraction imaging systems. It results from the interplay of fixed pattern noise or absorption inhomogeneities and misregistration between the two subtracted images.[8] For example, a new noise source is created from image intensifier structure noise in combination with TV camera time jitter. Figure 4 illustrates the noise-generating process. The $V(t)$ trace shows image intensifier fixed pattern noise along a single video line. In the absence of any time jitter between the first and second video images, these structural variations would subtract out in the final image without penalty. However, in the presence of some time jitter, represented by Δt, the image intensifier structure noise patterns no longer line up spatially, resulting in the generation of a residual noise pattern following subtraction of the two images. This represents an increasingly important noise source in digital time-mode subtraction imaging.

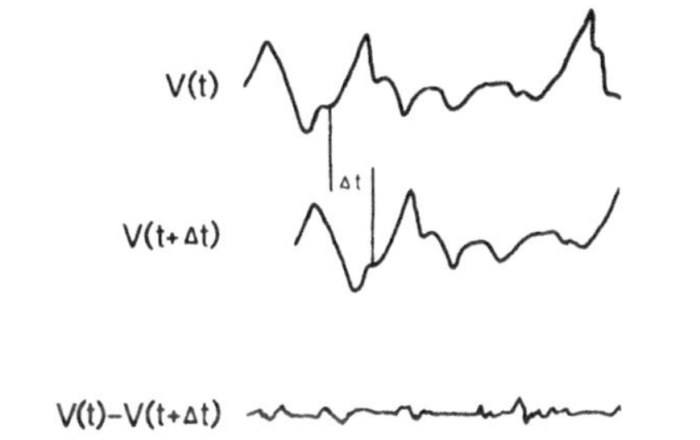

Figure 4. Noise-generating process from image intensifier structure noise (fixed pattern) and time jitter. Time shift of Δt results in incomplete subtraction of fixed-pattern noise, creating new noise source in subtraction imaging.[8]

A closely related source of noise results from various image reregistration processes. In this case, spatial shifting of the images before subtraction causes incomplete subtraction of image intensifier structure noise in much the same way that time jitter produces a residual noise pattern. An additional noise source results from the interplay of patient absorption variations in combination with patient motion. The normal variations across the patient in terms of X-ray absorption do not line up in the event of patient motion between the two video images. The subtracted image therefore shows a residual noise pattern due to the incomplete subtraction of the patient absorption inhomogeneities. Image reregistration in subtraction angiography may therefore lead to an increase in image noise in some cases although image resolution has been improved.

Scattered radiation is known to have a tremendous degradation effect on image quality in conventional film radiography.[15] Film requires a relatively fixed radiation fluence for a given film density, and is in general a contrast-limited imaging detector. The effects of scatter can be analyzed in a fairly straightforward way. The effect of scattered radiation in digital imaging, where image processing is applied, is not well understood. The scatter component is common to both before and after images in time-mode subtraction, and, on first thought, would appear to be removed without penalty in the subtraction process. Detection of the scattered radiation, however, can be shown to reduce the detail SNR and image gray levels, leading to the addition of noise and a loss of digitization accuracy.

3. CONSPICUITY AND IMAGE SUBTRACTION

The visibility of a diagnostic detail depends not only on the physical properties of the detail but also on the properties of the structures that surround it. The concept of lesion conspicuity was introduced by Kundel and Revesz[16] in an attempt to provide a quantitative description of this process. Conspicuity was defined as the ratio between lesion contrast and the complexity of surrounding anatomy. It was shown that the measure correlated well with the probability of detection of low-contrast detail in chest radiographs. The influence of image subtraction on lesion conspicuity can be appreciated by comparing the detection capabilities of conventional radiography and CT.

It is now generally recognized that conventional screen–film radiography results in the loss of a great deal of diagnostic information contained in the exit beam from the patient.[17] The development of CT scanning has made it possible to see soft tissue structures which cannot be visualized using conventional radiographic techniques. It is helpful to estimate the differences in detection capabilities of CT scanning and screen–film techniques. The density discrimination of CT scanners is usually specified at about 1% of the absorption coefficient of water.[18] The linear attenuation coefficient of water at 70 keV is 0.19 cm^{-1}. This means that CT scanners can image a voxel containing a homogeneous mixture of substance when the attenuation coefficient is either larger than 0.209 cm^{-1} or less than 0.179 cm^{-1}. In comparison, experimental measurements have shown that the minimum detectable subject contrast with screen–film systems is about 2%.[19,20] In order to compare the detection capabilities of these two systems, it is necessary to find a common parameter such as subject contrast. If the resolution of the CT scanner is assumed to be 0.5 cm, the equivalent subject contrast of a 1% difference in attenuation coefficient relative to water is about 0.1%. These results indicate that the contrast detection capabilities of these two systems differ by one order of magnitude for larger, nonresolution-limited details. Although several factors contribute to this difference in detection capability, the tomographic nature of the CT image represents one of the major advantages of CT systems. In conventional two-dimensional projection radiography, tissue inhomogeneities lying above or below a diagnostic detail may add background complexity, thus decreasing detail conspicuity. The ability to remove information not contained in the tomographic layer with CT can be considered a kind of image processing operation, in that a subset of the total information is selected for more precise display. Subtraction angiography provides similar image processing in that the total information content recorded in each image is reduced by the subtraction process. By subtracting images before and after the injection of iodine, all information except that related to blood vessel structure and dynamics is removed. Contrast examinations such as angiography represent a unique opportunity to perform this kind of image processing.

Film subtraction techniques are routinely used in angiographic examinations. The time required to carry out the photographic subtraction process and the associated difficulties of film registration limits its applications. A major advantage of digital radiography is the ability to subtract images rapidly and conveniently as well as to process the difference image for optimum visual presentation.

4. THEORETICAL ANALYSIS

Radiologic image formation is a statistical process involving the detection of large numbers of photons. The inherent limitation in X-ray imaging is therefore always set by the statistical noise.

4.1. Detail SNR

A fundamental parameter defining the available information content of a radiologic image is the detail SNR (SNR_d). It has been defined in several ways in the literature.[21–24] For the present discussion, a simplified presentation will be made. Let the radiologic detail be a disk object of area A imaged in a background of white noise. Coloring the noise, including the effects of source modulation transfer function (MTF), etc., will add complexity not required here. An incident radiation exposure to the patient contains some unknown photon fluence of N_0 per unit area. A fraction of those photons are transmitted through the patient and detail and are then available for detection at the image plane (N_d). The available photon fluence which has passed through an area adjacent to the detail can be similarly defined (N). In addition, there is a photon fluence of scattered photons (N_s) present at the image plane. The quality of the image is controlled by these differences in the number of transmitted photons. First, the expression of the SNR of primary radiation will be written. Scatter will be added later. The image signal is given by the quantity $|N - N_d|$. The total image in air incident on the detector is then given by the spatial distribution of these image signals.

The detail SNR is defined as the ratio of the image signal to the statistical noise associated with the image signal. The standard deviation of the photon fluences transmitted in each resolution element are equal to $N_d^{1/2}$ and $N^{1/2}$ and the total noise is then $(N_d + N)^{1/2}$. The detail SNR (SNR_p) at the image plane for primary radiation is then

$$SNR_p = |N - N_d|/(N + N_d)^{1/2} \tag{3}$$

For low-contrast objects ($N \approx N_d$), it can be shown that this expression reduces to

$$\mathrm{SNR}_p = (N/2)^{1/2}(1 - e^{-\Delta}) \tag{4}$$

where N is the number of transmitted photons per resolution element and Δ describes the X-ray attenuation properties of the detail and its surroundings for the beam spectrum transmitted through the patient and is given by

$$\Delta = t_d[(\mu/\rho)_d\rho_d - (\mu/\rho)\rho] \tag{5}$$

where t_d is the thickness of the detail, $(\mu/\rho)_d$ and μ/ρ are the mass attenuation coefficients of the detail and surrounding, ρ_d and ρ are the physical densities of detail and surrounding. Subject contrast can be defined for low-contrast details as

$$C_p = 1 - e^{-\Delta} \tag{6}$$

The important observation here is that subject contrast is a nonstochastic parameter, being a function only of the attenuation coefficients of the object and detail, and the photon spectra transmitted by the object.

The SNR and subject contrast are related (Eqs. 4 and 6):

$$\mathrm{SNR}_p = (N/2)^{1/2}\, C_p \tag{7}$$

The detail SNR, for a given detail located in a given object, therefore, can be increased by utilizing more photons to reduce noise or by modifying the exit beam spectrum to increase subject contrast.

Both the SNR and subject contrast are a function of scattered radiation. Scatter will add over the imaging plane, and is a function primarily of beam area and object thickness. A scatter degradation factor (SDF) can be defined[25] as $(1 + N_s/N)^{-1}$.

The detail SNR (SNR_s) in the presence of scatter is then

$$\mathrm{SNR}_s = \mathrm{SNR}_p(\mathrm{SDF})^{1/2} = \left(\frac{N}{2}\right)^{1/2} C_p \left(1 + \frac{N_s}{N}\right)^{-1/2} \tag{8}$$

The SNR is degraded in proportion to $(\mathrm{SDF})^{1/2}$ if we compare images with the same photon fluences. As is well known, scatter has a major effect on image quality, necessitating the use of scatter removal devices. The device most commonly used is a grid although more recent work by Barnes and others with scanning multiple slit assemblies has shown significant promise.[26] When a scatter removal device or technique (which can be an air-gap) is used, the scatter degradation factor (SDF_G) is

$$SDF_G = \left(1 + \frac{T_s N_s}{T_p N}\right)^{-1} \tag{9}$$

where T_s and T_p are the transmission factors for scatter and primary, respectively. For an ideal grid ($T_s = 0$ and $T_p = 1$), the SNR_G is equal to the SNR_p.

The detail SNR with a scatter removal device is then

$$SNR_G = (N/2)^{1/2} C_p (SDF_G T_p)^{1/2} \tag{10}$$

4.2. Detectability Threshold and Image Gray Levels

The ability to detect a given subject contrast requires that the SNR exceed a certain minimum value. Rose[27] and Strum and Morgan[28] have suggested a minimum value of 5, i.e.,

$$S/N \text{ (threshold)} \geq 5 \tag{11}$$

The threshold value is not known precisely, and in general is a function of several known and unknown factors. Values varying from 2 to 7 have been reported. The threshold value is meaningful only when additional noise added by the imaging system is negligible. Detection here means in a statistical sense at a given confidence level for an ideal detector. This means a given imaging system can detect a small contrast which is not visible to the radiologist in the final image display.

Using Eq. (10), the maximum SNR is

$$(S/N)_{max} = N^{1/2} (SDF_G T_p)^{1/2} \tag{12}$$

The minimum SNR required for detection is 5 (Eq. 11). The number of gray levels (G.L.) statistically present in the image can then be defined[21] as

$$\text{G.L.} = (\text{S/N})_{\text{max}}/(\text{S/N})_{\text{min}} = (1/5)N^{1/2}(\text{SDF}_{\text{G}}T_{\text{p}})^{1/2} \quad (13)$$

It is important to note that Eq. (13) applies to the output of a linear detector, phosphor screen, image intensifier input phosphor, and so on. It gives the theoretical number of gray levels present in the image before it is subjected to any readout device. The number of gray levels may be expressed in terms of the number of bits (b) per resolution element by converting to the binary system:

$$b = \ln(\text{G.L.})/\ln 2 \quad (14)$$

Equation (13) indicates that the number of gray levels increases with the square root of the transmitted exposure and is a strong function of the quantity of scatter present. The number of gray levels present in an image gives a useful index of the available image information content or the ability of an imaging system to discriminate between small image contrasts. The number of gray levels coded in the exit image from the patient may be much larger than those recorded by an actual imaging system or those visible to the eye in the final display.

4.3. Contrast–Detail Relationship for Threshold Detectability

The minimum X-ray fluence or exposure at the image plane which is required for the detection of a given signal with primary contrast C_{p} and with size (diameter) d can be determined from Eq. (10). This expression must first be converted from photon fluence per resolution element to photon fluence per detail size. Consider the detail to be square with side dimension of d. Then the number of photons contained in this detail area is Nd^2 where N is as defined in Eq. (3). Equation (10) can then be modified to

$$C_{\text{p}}^2 d^2 = 2(\text{SNR})^2\,(1/\text{SDF}_{\text{G}})\,T_{\text{p}} \quad (15)$$

The minimum values for contrast or detail size are lower bounds on detectability set by statistical noise where the threshold SNR is equal to 5. Detectability here does not necessarily mean the signal will be visible unless some image processing has been used. This kind of calculation further assumes the image recording system does not add noise during the detection or display process, and, further, that the radiologist will accept an image where quantum mottle is present.

4.4 Detector Quantum Efficiency

The detector quantum efficiency (DQE) is a measure of the efficiency of information transfer or extraction by an imaging component. Jones[29] defined the DQE as

$$\text{DQE} = (\text{SNR}_o)^2/(\text{SNR}_i)^2 \tag{16}$$

where SNR_o is the SNR of the output signal and SNR_i is the SNR of the input signal. An ideal detector or image transmission system would have a DQE of 1.0. Knowledge of the DQE of a real imaging system or component allows one to predict the maximum improvement possible if the system were replaced with an ideal component or system. A component can be considerd as a black box where the DQE describes its efficiency of SNR transfer. In this regard it has pleasing similarities to the MTF. Wagner[17] has used the DQE approach to describe scattered radiation. The real value of the DQE is as a descriptor of various components in the imaging chain. The DQE of various stages of information transfer in an imaging system can be calculated independently to aid in the isolation of the weak component.

4.5. Minimum Patient Exposure for Detection

The above theoretical analysis has been based on radiation exposure or photon fluence at the imaging plane. In most cases this simplifies the discussion, since the patient exposure can readily be determined when the exit exposure is known. The number of incident X-rays to the patient (N_0) is given by $N_0 = Ne^{\mu\rho t}$ where μ, ρ, and t are the mass attenuation coefficient, density, and thickness of the patient. The entrance patient exposure (X) is then given by

$$X = Ne^{\mu\rho t}K(E)B \tag{17}$$

where $K(E)$ is the well-known factor to convert from photon fluence to air exposure as a function of photon energy and B is the backscatter factor. The SNR (Eq. 10) expressed in terms of patient exposure is then

$$\text{SNR}_\text{G} = \frac{Xe^{-\mu\rho t}}{2K(E)B}\, C_\text{p}(\text{DQE}_\text{G})^{1/2} \tag{18}$$

This expression can be used to determine the SNR as a function of, for example, photon energy or patient thickness. Alternatively, it may be used to determine the minimum patient exposure required for detection of a given detail as a function of photon energy.

4.6. Sample Calculations for Digital Angiography

Using the set of equations developed above, it will be instructive to carry out a sample set of calculations for digital angiography. Let us assume that the dilution factor for IV arm injections of contrast media will be about 5% for most organ systems. This quantity is expected to vary widely from patient to patient, the particular organ site, and the injection conditions. The iodine concentration of Renografin-76 is 370 mg/cm^3. Using these quantities, the effective thickness of iodine in a 1-mm-diameter vessel with IV injection is 1.85 mg/cm^2. This quantity will be used for all calculations given below.

Calculated detail SNRs at 40 keV for the iodine detail described above are shown in Figure 5. It can be seen that about 0.5 mR/frame is required for statistical detection in the presence of a scatter-to-primary ratio of 1

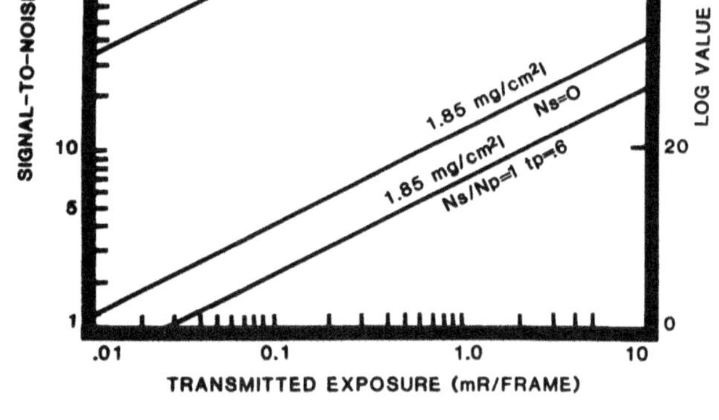

Figure 5. Signal-to-noise ratios at 40 keV for 1.85 mg/cm^2 iodine detail. An exit exposure rate of 0.5 mR/frame is required for statistical detection in the presence of a scatter-to-primary ratio of 1.0 and primary transmission of 0.6. Without scattered radiation, the required exposure is reduced to about 0.15 mR. The maximum SNR (top curve) with an exit exposure of 0.6 mR/frame is about 250:1. The right axis shows SNRs in dB.[7]

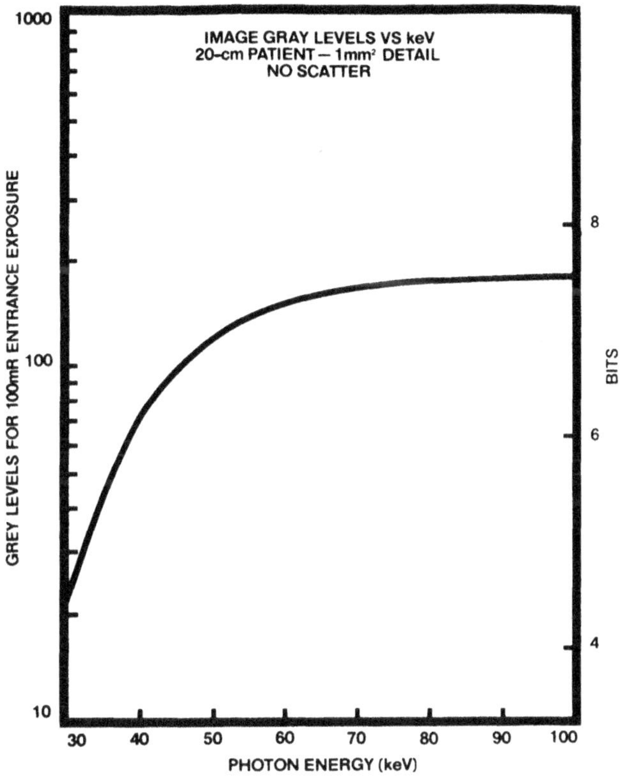

Figure 6. Number of gray levels or bits as a function of photon energy for patient entrance exposure of 100 mR (20-cm-thick patient, no scatter).[7]

and a grid transmission factor for primary radiation of 0.6. Without scattered radiation, the required exposure is reduced to about 0.15 mR. Also shown in Figure 5 (top) is the maximum SNR versus transmitted exposure. The maximum SNR for 1 mR at 40 keV is seen to be 300. If the minimum SNR for detectability is 5, the dynamic range of the exit beam is then 60.

Figure 6 shows the number of gray levels or bits as a function of photon energy for an entrance exposure of 100 mR to a 20-cm-thick patient (no scatter). The number of gray levels increases with increased photon energy due to increased penetration at the higher beam qualities. Beam energies beyond 60 to 70 keV show little improvements in gray level. It is of interest to note that an exit exposure of 8 mR is required to achieve 8 bits of gray scale.

Figure 7 shows the minimum patient entrance exposure required for

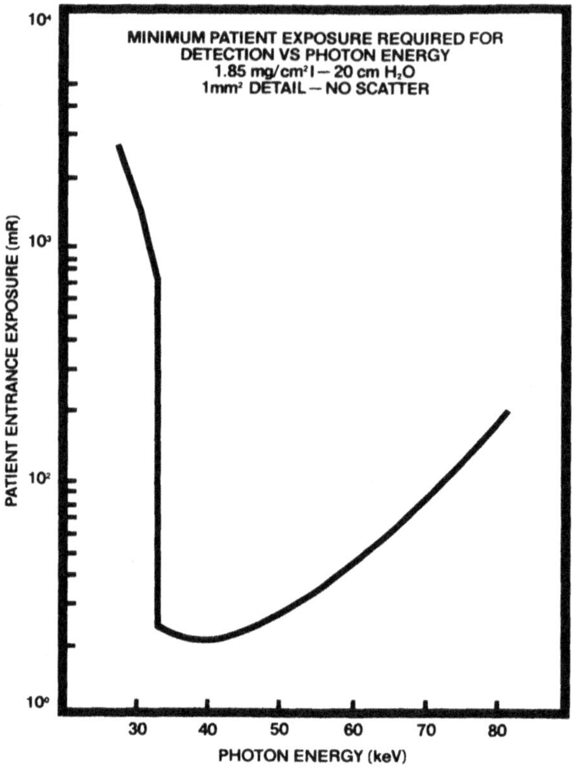

Figure 7. The minimum patient entrance exposure required for statistical detection of a 1-mm² detail of 1.85 mg/cm² iodine thickness as a function of photon energy.[7]

detection of a 1-mm² detail (1.85 mg/cm² of iodine, 20-cm patient thickness) as a function of photon energy. An idealized case without scatter is assumed. The reduction in required patient exposure at the iodine *K*-edge is dramatic as it varies with the square of primary contrast. A second important observation concerns the relationship of required exposure for energies above the *K*-edge of iodine. The required exposure does not immediately increase with higher photon energies but actually shows further decreases up to about 40 keV. It can also be seen that 50-keV photons are about as efficient as 33-keV photons in terms of patient dose required for detection. This curve has important implications concerning the desirability of achieving monoenergetic X-ray beams. A beam spectrum covering the energy range 33–50 keV would appear to be as effective as a monoenergetic beam positioned just above the *K*-edge of iodine.

4.7. Summary of Theoretical Analysis

The X-ray beam which exits from the patient carries all of the diagnostic information. All processes which occur subsequent to this stage (detection, amplification, display, and so on) serve only to degrade image information. No new information can be added. The above theoretical analysis has used image information theory to quantify the information carried by the exit beam under various conditions. This represents a new approach to designing X-ray imaging systems. Before a system is engineered to extract and display image information, we can first quantify and characterize the information available for extraction. The results presented above cannot of course be used to predict the final image quality without first quantifying the effect of each component of the image acquisition system on the final image. It is believed that this type of analysis can aid in future designs of imaging systems and, in particular, digital radiography systems.

5. EXPERIMENTAL MEASUREMENTS OF NOISE

The experimental results presented here were obtained from measurements on an image intensifier–TV camera digital system operating with 512^2 pixel resolution, 9-bit digitization, and 8-bit image storage. The results are therefore more or less representative of digital video systems at the time of this writing.

5.1. SNR

The SNR of a digital image is usually defined as the maximum signal size divided by the standard deviation of the noise fluctuations. (For example, with an 8-bit image memory, the maximum signal is 255 gray levels in the computer.) This is a typical method of expressing the SNR but can be misleading without a common reference point. For example, with this method a 10-bit system would have a fourfold higher SNR than an 8-bit computer, even when both showed the same absolute standard deviations of image noise. This has led some investigators to express the noise properties of the system by using the detail SNR (SNR_d).[30] The signal is here defined as the differential signal size of some defined object or detail contained within another, larger homogeneous object. For example, the detail might be a 5 mg/cm^2 thick iodine disk located in a surrounding plastic phantom. The SNR_d gives a parameter more representative of the potential of the imaging system to detect low-contrast objects. Since the SNR_d is independent of stated imaging system specifications, it provides a universal parameter which can be more accurately related to image detectability and system performance.

There are several methods available to measure the SNR depending on where it is to be measured and whether SNR_{max} or SNR_d is required. It may also be desirable to measure the SNR of the video imaging chain before a digital system is added. This technique requires the use of an oscilloscope, preferably with dual channels. Peak-to-peak noise can be determined directly by observing the scope trace. For improved accuracy the tangential method can be used. The underlying theory is elegant and the technique is easily used by anyone who has a dual-channel scope. The procedure is to connect the noise source to both channels of the scope having alternate-sweep capability. With both channels identically calibrated, one adjusts the voltage offset until the dark band between the two noise traces just disappears. The separation of the two noise-free baseline traces after the signal is removed represents 2 times the rms noise. This technique is based on the fact that two identical Gaussian-distribution curves combine to form a single smooth curve without a dip when they are separated by 2 times the standard deviation. The technique has a reported accuracy of 1 dB for measuring rms noise.

The SNR of a digital imaging system can be determined directly from image data digitally recorded in the computer.[8] In the case of an image intensifier–TV imaging detector, it is desirable to perform the calculations on subtracted images to remove shading and structure noise sources. It must be remembered, however, that the structure noise has been removed, especially in the absence of time jitter. The calculations consist of selecting an area of interest in the image from which the mean and standard deviation are computed. The SNR is defined here as the peak signal, corresponding to 255 gray levels in an 8-bit computer, divided by the standard deviation of the noise. Since this SNR is calculated from the image recorded in the computer, all other noise sources of the system after the computer are excluded. Noise generated in the DAC and final TV monitor, which are believed to be insignificant, are of course excluded.

One of the first observations to be encountered in measuring the SNR is the relatively large variations in the experimental results which are typically obtained (see Figure 8). The rms noise and the mean signals were measured separately for a typical subtraction run. The results indicate that the signal fluctuates with a standard deviation of approximately 0.2% while the noise varies by 4.3% for a series of 15 images. On repeated trials we would theoretically expect the SNR to be asymptotically normally distributed with a standard deviation of about 10%, in excellent

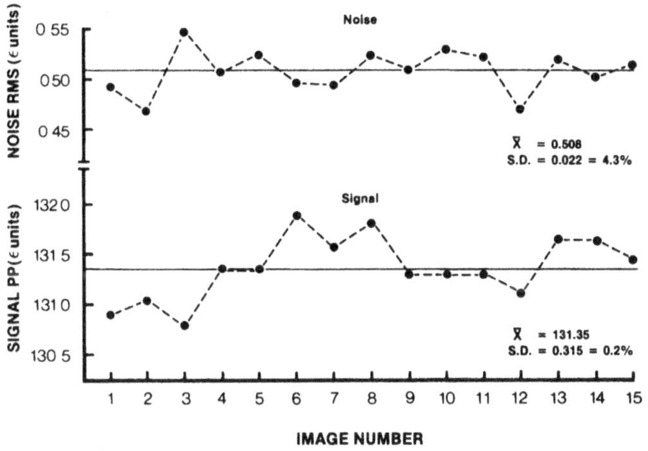

Figure 8. Variations in measured signal and noise levels during series of subtraction images.[8]

agreement with the results shown here. Experimentally measured SNRs can therefore be expected to vary significantly due to statistical variations.

Figure 9 shows the effect of frame averaging on the measured SNR of subtracted images. With the image intensifier exposed to an input exposure level of 0.2 mR/frame, subtraction images were obtained with averages of 1, 2, and 4 TV frames/image. The solid line shows the expected

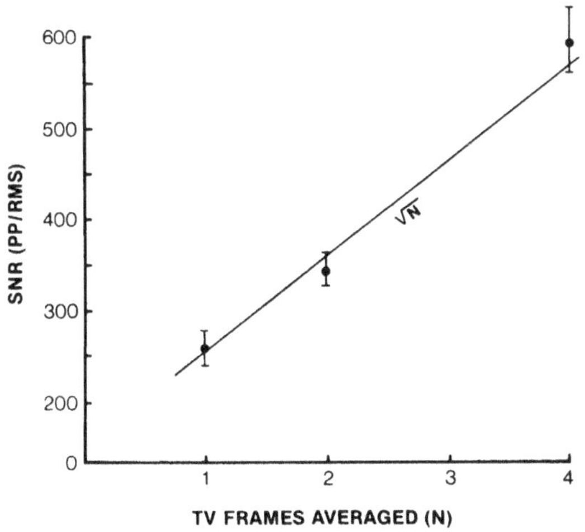

Figure 9. Measured SNR as a function of number of frames averaged per subtracted image.[8]

$\sqrt{N}$ relationship, indicating that the experimental data agree with expected results based on random fluctuations. This technique can be used to verify the performance of the measurement technique or to determine if structure noise is present, i.e., frame averaging would not be expected to reduce fixed pattern noise.

Quantum noise is well recognized as a fundamental determinant of SNR in all imaging systems. In order to determine the optimum set of operating conditions in the clinical setting, it is of significant value to know how the SNR is related to radiation exposure level. This determines the best exposure techniques for image quality while minimizing radiation dose to the patient. The total system SNR can be measured as a function of input radiation exposure to the image intensifier. Such results, shown in Figure 10, indicate a total system SNR of 750:1, with an exposure level of 1 mR/frame at the detector. If other significant noise sources had been present, the total SNR would have leveled off at a lower input radiation dose.

5.2. Scattered Radiation and Detail SNR

The effect of scattered radiation on image quality in subtraction radiography has been discussed previously. A phantom was used to generate scattered radiation, and then subtracted images were acquired in the computer, i.e., a mask was obtained, a small aluminum disk was placed on the phantom, and a second image was taken and subtracted. The detail

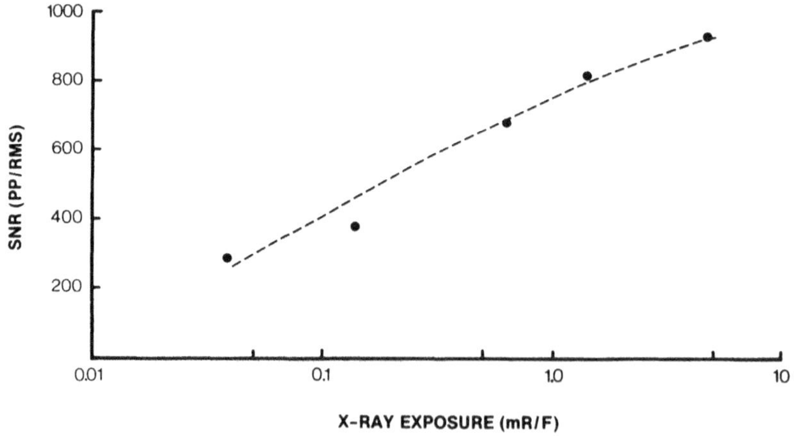

Figure 10. Measured system SNR versus radiation exposure level at image intensifier input (6-in. II, 1 frame).[8]

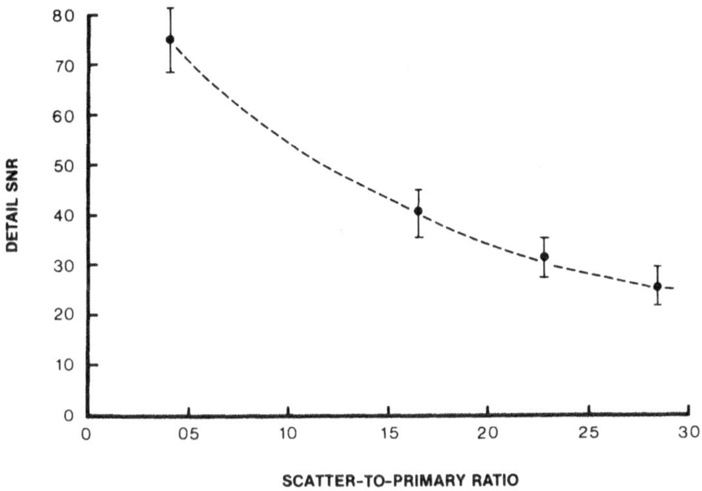

Figure 11. Measured detail SNR as a function of scattered radiation level. Dashed line shows expected response of $(1 + N_s/N)^{-1}$.

SNR was calculated from an area of interest behind the aluminum disk and from an area beside the disk image. The signal is defined as the difference between the average values obtained within these two areas. The noise is the rms fluctuations in the area beside the disk.

The measurements were repeated for varying levels of scattered radiation. The amount of scatter was varied by changing the field size while keeping all other parameters of the system constant. Figure 11 shows that the detail SNR decreases with increasing levels of scattered radiation. These results indicate that the detail SNR varies approximately as (1 + SDF).

5.3. *Wiener Power Spectra*

The noise power (Wiener) spectrum provides an analysis of the noise variances into their spatial frequency components. It therefore specifies the amount of noise variance per unit spatial frequency interval. This approach has all of the advantages of Fourier analysis, i.e., the ability to cascade the effects of specific imaging components of a system to obtain the overall system noise properties.

The experimental noise power spectrum reported here was determined from digital data recorded in the computer. Each image (512^2 pixels, 8 bits) used 1024 data points per calculation. A detrending polynomial of degree 0 was fit by least square techniques. A Henning correlation window of 8 pixels was used. Each of the 16 autocovariance functions was averaged. The Wiener power spectra were then calculated from the Fourier transform of the averaged autocovariance functions. One of the limitations of analyzing data already digitized and recorded in the computer is that these data cannot be sampled at a rate greater than 512 pixels/line. The technique is therefore limited to analyzing spatial frequencies up to only one-half the sampling rate, the Nyquist sampling theorem. Therefore, bandwidth limits set by the computer and the TV camera cannot easily be separated.

The autocovariance functions, used to calculate the power spectra, can also be analyzed in the spatial domain. They can be used to characterize the correlation between neighboring points or pixels in a statistical manner. For example, a purely random function will go to zero with a lag of one shift while systems with repetitive frequency components will show correlations at shifts corresponding to their base frequency. Figure 12 shows calculations of the autocovariance functions for three images obtained at three different exposure levels. It can be seen that the noise is correlated over distances varying from about 1 to 3 pixel widths. When the exposure is increased from 0.035 mR/image up to 3.5 mR/image, the correlation of data reduces sharply, such that at the higher exposure level there is little correlation of the data, even at 1-pixel shifts. This indicates that the noise is uncolored by the imaging system at the higher exposure

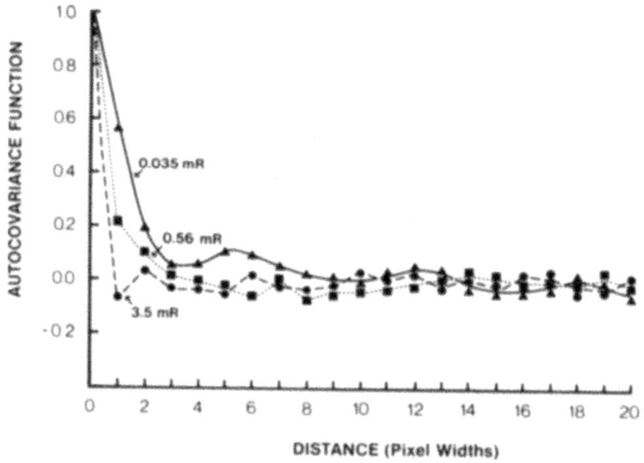

Figure 12. Autocovariance functions at three input radiation exposure levels for a digital video subtraction radiography system.

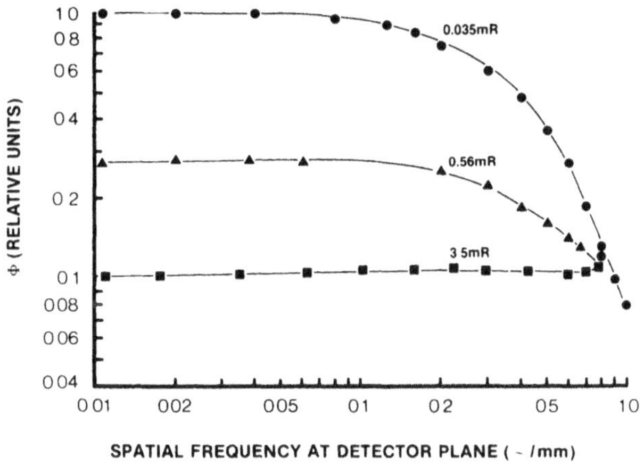

Figure 13. Wiener power spectra of three digital subtraction images taken at three input radiation exposure levels (subtraction, 9-in. II.)

level, i.e., the imaging chain is not a major factor in limiting spatial frequencies in the final image for this system.

The noise power spectra for the three exposure conditions of Figure 12 are shown in Figure 13. The vertical axis has been normalized to the low exposure condition, thus providing a relative measure of the variance. As would be anticipated from Figure 12, the frequency content as well as the magnitude of the noise varies with the exposure level. It can be seen that the noise amplitude is reduced at all frequencies when the exposure is increased and that the frequency contents also shift. At 0.035 mR/frame, quantum noise dominates, and the shape of the power spectrum is believed to be dominated by the pixel spacing in the computer. When the exposure is increased to 0.56 mR/frame, TV camera electronic noise makes an important contribution to the total noise. At the high exposure level of 3.5 mR/frame, quantum noise has a small contribution to the total noise. Since the frequency content of the noise of the 3.5 mR/frame image is very flat, this would suggest that quantization noise is the dominant noise source since it cannot be bandwidth limited due to the experimental technique.

The measured Wiener power spectra for three input modes of the image intensifier are shown in Figure 14. A constant input exposure of 0.12 mR/frame was used to obtain subtracted images. The results indicate that the noise power is reduced when the field size is increased while maintaining a constant-input exposure. This is consistent with expected

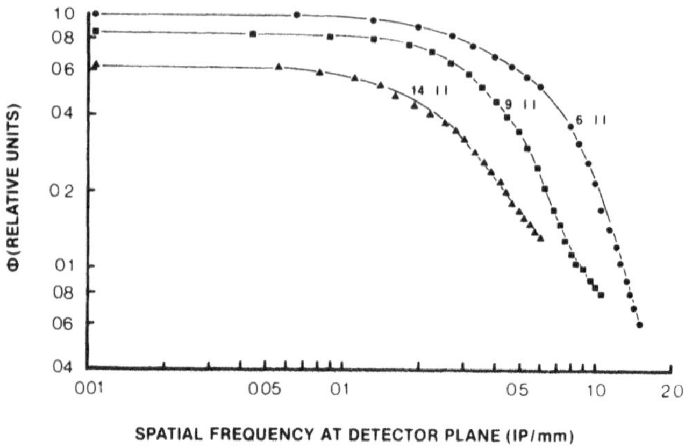

Figure 14. Wiener power spectra of three subtracted images taken with a digital video imaging system at three image intensifier format sizes (0.1 mR, 1-frame images).

results since a larger quantity of X-ray photons are collected with the larger field sizes, and in all cases the image is recorded as a 512 × 512 pixel image. The spatial frequency is defined at the input plane of the image intensifier in all three measurements. It can be seen that the bandwidth is reduced as the image intensifier format size is increased. The shape of the rolloff of the 6- and 9-inch modes suggests that the computer pixel spacing plays a dominant role in determining spatial resolution. The shape of the curve is significantly different for the 14-inch mode of the image intensifier, suggesting that the rolloff of the curve is dominated by the bandwidth of the TV camera. This result, supported by calculations, indicates that the 5-MHz bandwidth of this TV is not adequate to fully extract the image data available in the 14-inch mode.

5.4. Phantom Tests of Iodine Detectability

A digital test phantom has been designed to provide performance evaluations of digital radiography systems.[30] Iodine details which are stable over long time periods permit measurements on iodine detectability of the imaging system in a reproducible manner. The phantom is constructed from approximated tissue equivalent material, and contains a variety of test objects designed to evaluate specific properties of digital imaging systems. A new chemical mixture was developed to provide a stable complex containing elemental iodine bound to a low-Z molecule.[31] This new mixture allows fabrication of iodine details with tissuelike characteristics which contain small iodine concentrations representative of

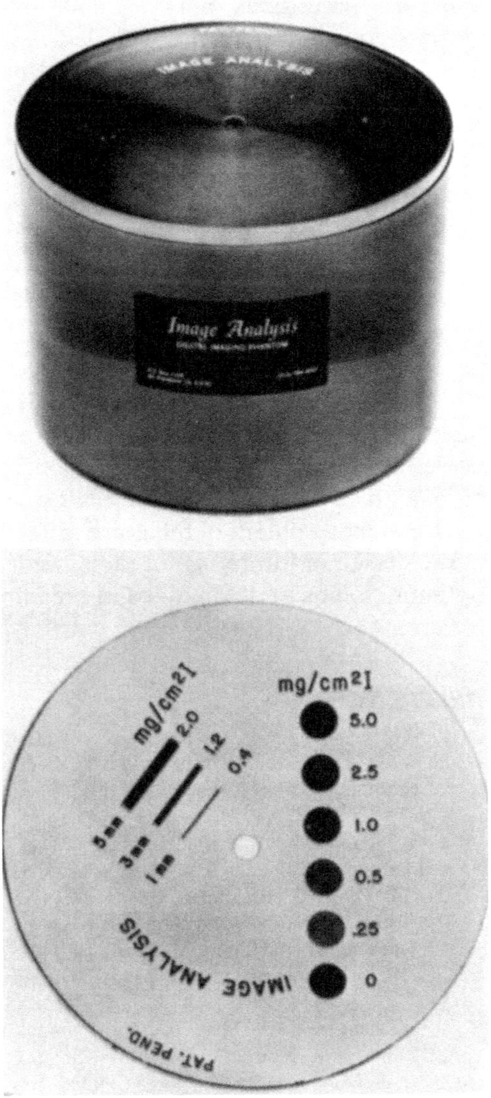

Figure 15. Digital radiography test phantom with iodine test plate: varying concentrations of iodine in test details allows measurement of iodine detectability and detail SNR.[30]

clinical angiography. Five test disks of 1.5-cm diameter are incorporated into the iodine test plate. Effective iodine thicknesses of 5, 2.5, 1.0, 0.5, and 0.25 mg/cm^2 of iodine are provided. The details are sufficiently large that their detection is determined primarily by the noise properties of the imaging system and are weakly dependent on its spatial resolution. In addition, three tubes of 1-, 3-, and 5-mm diameter containing iodine of various concentrations are included.

The test phantom and iodine test plate are shown in Figure 15. Use of the phantom with digital video subtraction angiography systems has produced SNRs varying from 300:1 to 700:1 for typical exposure levels of 0.5–1.0 mR/image at the detector. Iodine detectability of 0.5 mg/cm^2 at an exposure of 1 mR/image is typical and in good agreement with the theoretical calculations presented above.

6. SUMMARY

The sources of noise in digital radiography systems have been identified and analyzed. Quantization error noise and time jitter–structure noise are discussed as new noise sources of digital systems. Noise can be quantified by the use of the SNR and by Wiener power spectra when the frequency content of the noise is needed. Noise analysis is useful both in the design of future digital radiography systems and in the selection of optimum exposure techniques in the clinical environment.

REFERENCES

1. R. A. Kruger, C. A. Mistretta, and S. J. Riederer, Physical and technical considerations of computerized fluoroscopic difference imaging, *IEEE Trans. Nucl. Sci.* **28,** 205–212 (1981).
2. S. J. Riederer, F. A. DiBianca, and J.-P. J. Georges, Performance characteristics of a digital fluorographic system, *Proc. SPIE* **273,** 88–95 (1981).
3. M. M. Frost, S. Nudelman, and T. W. Ovitt, Digital acquisition system for photoelectronic radiology: A performance overview, *Proc. SPIE* **233,** 54–60 (1980).
4. H. Roehrig, S. Nudelman, M. P. Capp, and M. M. Frost, X-ray image intensifier video system for diagnostic radiology. Part I. Design characteristics, *Proc. SPIE* **127,** 216–225 (1977).
5. R. A. Kruger, C. A. Mistretta, and J. Lancaster, A digital video image processor for real time x-ray subtraction imaging, *Opt. Eng.* **17,** 652–657 (1978).
6. P. O. Scheibe and A. J. Thomas, Noise sources in digital fluorography, *Proc. SPIE* **314,** 202–210 (1981).
7. B. A. Arnold, D. Borger, P. Scheibe, and J. Pawlicki, Digital Video Subtraction Angiography, *Scientific Exhibit, Radiological Society of North America,* Chicago (1980).
8. B. A. Arnold and P. O. Scheibe, Noise analysis of a digital radiography system, *Am. J. Roentgenol.* **143,** 609–613 (1984). Vol 142
9. N. J. Pelc, Statistical aspects of digital x-ray imaging, in *Electronic Imaging in Medicine* (G. D. Fullerton, ed.), American Institute of Physics, New York (1984).
10. R. J. Gould, M. J. Lipton, P. Mengers, and R. Dahlberg, Digital subtraction fluoroscopic system with tandem video processing units, *Proc. SPIE* **273,** 125–131 (1981).

11. W. J. McIntyre, W. Pavlicek, and J. H. Gallagher, Imaging capability of an experimental digital subtraction angiography unit, *Radiology* **139,** 307–313 (1981).
12. B. A. Arnold, H. Eisenberg, D. Borger, and A. Metherell, Digital radiography—An overview, *Proc. SPIE* **273,** 215–226 (1981).
13. B. A. Arnold, H. Eisenberg, and D. Borger, Digital video angiography system evaluation, *Appl. Radiol.* **10,** 81–90 (1981).
14. B. Widrow, A study of rough amplitude quantization by means of Nyquist sampling theory, *IEEE Trans. Circuit Theory* **3**(4), 266–276 (1956).
15. G. T. Barnes and I. A. Brezovich, The intensity of scattered radiation in mammography, *Radiology* **126,** 243–247 (1978).
16. H. L. Kundel and G. Revesz, Lesion conspicuity, structured noise, and film reader error, *Am. J. Roentgenol.* **126,** 1233–1238 (1976).
17. R. F. Wagner, Noise equivalent parameters in general radiography: The present picture and future picture, *Photogr. Sci. Eng.* **21,** 252–262 (1977).
18. R. A. Brooks and G. DiChiro, Principles of computer assisted tomography (CAT) in radiographic and radioisotope imaging, *Phys. Med. Biol.* **21,** 689–732 (1972).
19. M. D. Ter-Pogossian, M. E. Phelps, E. J. Hoffman, and J. O. Eichling, The extraction of the yet unused wealth of information in diagnostic radiology, *Radiology* **113,** 515–520 (1974).
20. G. Cohen, L. K. Wagner, S. R. Amtey, and F. A. DiBianca, Contrast-detail dose and dose efficiency analysis of a scanned digital and a screen-film-grid radiographic system, *Med. Phys.* **8,** 358–367 (1981).
21. J. W. Motz and M. Danos, Image information content and patient exposure, *Med. Phys.* **5,** 3–22 (1978).
22. R. F. Wagner, Decision theory and the detail signal-to-noise of Otto Schade, *Photogr. Sci. Eng.* **22,** 41–46 (1978).
23. O. H. Schade, *Imaging Quality: A Comparison of Photographic and Television Systems,* RCA Laboratories, Princeton, N.J. (1975).
24. A. Rose, *Vision: Human and Electronic,* Plenum Press, New York (1973).
25. R. F. Wagner, G. T. Barnes, and B. S. Askins, Effects of reduced scatter on radiographic information content and patient exposure: A quantitative demonstration, *Med. Phys.* **7,** 13–18 (1980).
26. G. T. Barnes, H. M. D. Cleare, and I. A. Brezovich, Reduction of scatter in diagnostic radiology by means of a scanning multiple slit assembly, *Radiology* **120,** 691 (1976).
27. A Rose, The sensitivity performance of the human eye on absolute scale, *J. Opt. Soc. Am.* **38,** 195 (1948).
28. R. E. Strum and R. H. Morgan, Screen intensification systems and their limitation, *Am. J. Roentgenol.* **62,** 617 (1949).
29. R. C. Jones, Method of measuring the granularity of photographic materials *Photogr. Sci. Eng.* **2,** 57 (1958).
30. B. A. Arnold, Digital radiography test phantom, *Scientific Exhibit, Radiological Society of North America,* Chicago (1982).
31. B. A. Arnold, Digital test phantom and method for its fabrication, U.S. Patent Application 6,555,608 (1983).

5

Quantitative Aspects of Image Intensifier–Television-Based Digital X-Ray Imaging

O. Nalcioglu, W. W. Roeck, J. A. Seibert, A. V. Lando, J. M. Tobis, and W. L. Henry

1. INTRODUCTION

The utilization of video systems in diagnostic X-ray imaging has been around for quite some time. In addition to the visual diagnostic value of images obtained by such systems, the X-ray video systems also offer the potential for extracting quantitative information from the images. In this chapter we will not deal with quantitative determinations based on distance measurements. Even though the quantitation potential has been under investigation for the past decade,[1–3] it was not until the introduction of digital subtraction angiography (DSA)[4] that it received a concentrated effort. DSA makes it possible to visualize a small amount of radiopaque contrast agent within the vessels by means of various digital image enhancement methods which are discussed elsewhere in this book (Chapters 3, 7). Image processing takes place after the digitization of the video signal. Many users may assume that since the output is digital format, it must

O. Nalcioglu, W. W. Roeck, and A. V. Lando • Department of Radiological Sciences, University of California, Irvine, California 92716. Present address for A.V.L.: Philips Medical Systems, Inc., Shelton, Connecticut 06484. ***J. A. Seibert*** • Department of Radiology, University of California, Davis, California 95817. ***J. M. Tobis and W. L. Henry*** • Department of Medicine (Cardiology), University of California, Irvine, California 92716.

be as accurate as the data in computed tomography (CT) systems. Unfortunately, this is not true, and the reason is the existence of some basic differences between the CT and image intensifier–television (IT–TV)-based digital X-ray imaging systems. In digital fluoroscopy (or radiography), one usually uses large-area detectors such as an X-ray II. Due to poor detection geometry, the scattered photons within the patient seriously degrade the information obtained by a digital fluoroscopic system. In addition to poor geometry, the II detector itself is the source of another problem. This is the lateral propagation of light within the II, which is known as the veiling glare.[5] Veiling glare causes nonlinear upward shift of the video levels in the dark parts of an image. A final major problem is the beam hardening within the patient. The source of this problem is the preferential absorption of low-energy X-rays by an object, and the variation of effective beam energy as a function of object thickness. All these problems result in an overestimation of measured digital fluoroscopic numbers, but an underestimation of quantities which are based on the X-ray projection. Even though the three physical problems mentioned above reduce the accuracy of absolute measurements considerably, we have found in certain relative measurements that one may have a cancellation of errors.[6] However, the success in the computation of these parameters should not give a false sense of confidence. The development of correction techniques to reduce these errors is possible and some of them will be discussed in this chapter. Our basic philosophy differs substantially from many researchers who have attacked the same problems. We are not interested in applying empirical "Band-Aid" correction techniques which may only be applicable in a limited situation. We will discuss general methods which are applicable to broader phenomena, and, specifically, will improve the accuracy of digital fluoroscopic numbers for absolute measurements.

In Section 2, we will review the imaging system and various components. Section 3 will cover a detailed discussion of the characterization of the three physical problems mentioned earlier. Effect of physical degradation factors on the measurement of volumes is reviewed in Section 4. Techniques to reduce the effects of these problems will be presented in Section 5. Some examples of typical applications will be discussed in Section 6. A summary will be given in Section 7.

2. SYSTEM DESCRIPTION

Most of the imaging systems used in digital fluoroscopy utilize an II–TV camera as the receptor. A typical X-ray video system used in diagnostic imaging is shown in Figure 1. The various components of the imaging chain are the X-ray generator, X-ray tube, II, TV camera, video processor, viewing and recording devices. In this chapter, we will discuss the propagation of a signal through the system and relevant equations describing this process. We will not be concerned with recording or viewing equipment.

2.1. X-Ray Generator and Tube

The term *generator* applies to that part of the system which controls and provides the high voltage and electrical power required to energize the X-ray tube. In quantitative measurements using an X-ray video system, it is essential that the X-ray tube output remains constant throughout the whole study. Pulse-to-pulse fluctuations in the X-ray fluence can be minimized by the use of 3-phase 12-pulse X-ray generators. For such generators, the maximum fluctuation in the tube output is about 3.5% of the peak voltage.

The X-ray tube should have a heat storage capacity of 300,000 heat units or more, so that a large flux of X-rays can be sent through the object. Another important aspect of the X-ray tube is the size and intensity distribution of the focal spot. The smaller the size, the smaller the objects we can image. Unfortunately, small focal spot size also implies loading a large amount of heat into a smaller area of anode, thus causing further X-ray flux limitations. For most applications, a 0.3/1.3-mm switchable focal spot is sufficient.

The function which describes the incident X-ray fluence at a location (x',y') in front of the object is given by

$$\text{Incident fluence} = N_0(x',y':E) \tag{1}$$

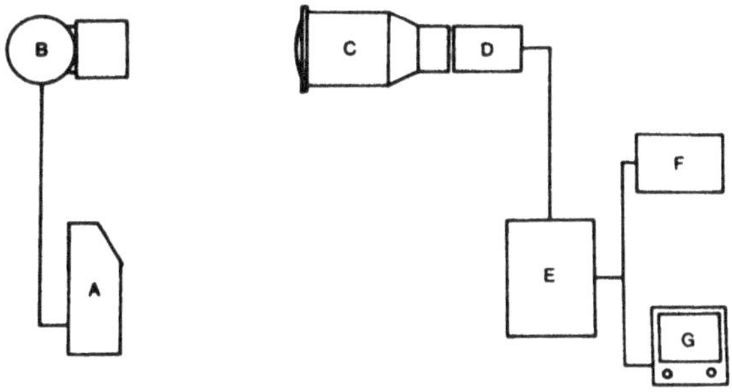

Figure 1. Overall imaging system: (A) X-ray generator, (B) X-ray tube, (C) image intensifier, (D) TV camera, (E) video image processor, (F) recording devices, and (G) monitor.

In writing Eq. (1), we assumed a parallel X-ray beam geometry and hence no dependence on z' where z' is the distance measured from the source. The energy variable E in Eq. (1) indicates the polychromatic nature of X-rays used explicitly.

2.2. Object

Let us assume that a three-dimensional object with a linear attenuation coefficient distribution $\mu(x,y,z:E)$ is placed into the X-ray beam. If the X-rays are emitted from the source in the z' direction, then the transmitted X-ray fluence is given by

$$N(x',y':E) = P(x',y':E) + S(x',y':E) \tag{2}$$

where S and P are the scatter and primary components, respectively. The primary X-ray distribution is related to the incident one by the following equation:

$$P(x',y':E) = N_0(x',y':E) \exp\left[-\int \mu(x',y',z':E)dz'\right] \tag{3}$$

where N_0 is the incident flux at location (x',y') and μ is the distribution of energy-dependent linear attenuation coefficients along the X-ray beam which is assumed to be parallel to the z' axis. The scatter component is related to the incident flux by a complex relationship. It is a complicated function of many variables including X-ray beam energy, object composition, and imaging geometry. A more detailed discussion of the scatter component will be presented in a later section.

2.3. Image Intensifier

All of the elements of an II tube are enclosed within a glass bottle that is highly evacuated. The input end of the tube has a coating consisting of the input phosphor, a thin transparent membrane, and the photocathode. The input phosphor is usually made of cesium iodide which absorbs the incident X-ray beam and emits visible light. Most IIs in use today have a cesium iodide input phosphor with a thickness of 100 mg/cm^2. The input phosphor thickness determines the detection efficiency as well as the spatial resolution capabilities of the II. The phosphor thickness is chosen as a trade-off between detection efficiency and spatial resolution. The limiting spatial resolution of a typical 9-inch II is about 4 lp/mm at the input phosphor. The detection efficiency is highly energy dependent. The visible light emitted from the phosphor passes through the thin membrane and is absorbed in the photocathode. As a result of photoelectric absorption in the cathode, electrons are released to the interior of the II tube. These electrons are accelerated and focused toward the output phosphor which is much smaller in size than the input one. The output phosphor absorbs the kinetic energy of the electrons and produces an intensified visible light. A fraction of electrons and produced light photons scatter within the II, thus causing a degradation of image quality known as veiling glare. In regions of low transmission adjacent to regions of high transmission of the detected X-rays, electrons and light produced from the brighter areas spread into the darker areas. This nonuniform added bias affects the contrast response as well as the quantitative integrity of the information contained in the two-dimensional projection image. In addition to veiling glare, there might be spatial distortions due to poor electron optics.

The generation of light images by an II may be described by the following general equation:

$$I(x,y) = \alpha \int\int D(x',y':E)T(x',y:x',y')N(x',y':E)dx'dy'dE \quad (4)$$

where α is a constant of proportionality and $D(x',y':E)$ is the detection efficiency of the II at point (x',y') for an X-ray of energy E. The function $N(x',y':E)$, which was defined in Eq. (2), indicates the transmitted X-ray fluence of energy E at a location (x',y'). The transfer function T takes an X-ray absorption event from a location (x',y') on the input phosphor and generates light at location (x,y) at the output phosphor. In defining T, we assumed that it is independent of energy E, but the energy dependence could also be included in a straightforward manner. Equation (4) was originally proposed by Kruger.[7] The transfer function T includes the various image degradation processes such as veiling glare and spatial distortions.

2.4. Television Camera

The TV camera converts the light flux incident on its target surface into an electronic signal which will eventually be digitized. Most of the TV cameras used in digital fluoroscopy utilize a lead oxide (plumbicon) target. It is absolutely necessary to measure and know the response of the TV camera for quantitative applications. In a later section, we will discuss how this is done. The conversion of the light image $I(x,y)$ into a video image by the camera is described by

$$V(x,y) = f[I(x,y)] \quad (5)$$

where I is defined in Eq. (4) and $f[I]$ is the transfer function for the process. In Eq. (5), $V(x,y)$ is the amplitude of the video signal at a location (x,y) within the image. Even though a general response function f is given in Eq. (5), in most cases it can be approximated by

$$f[I] = aI^{\gamma} \tag{6}$$

where a is a scaling factor and γ is a constant. For a plumbicon camera $\gamma \approx 1.0$.

2.5. Analog-to-Digital Converter

The analog-to-digital converter (ADC) is the device which converts the analog video signal into digital numbers. Due to the finite number of levels used, the ADC always adds additional noise (quantizing noise) to the signal. If the analog noise is normally distributed with a standard deviation σ, then an equation which relates the rms noise before and after the digitization can be derived. This relation is given by

$$\sigma_D/\sigma = 0.5 \left[\sum_{i=0}^{\infty} (2i + 1)^2 g_i(\epsilon/\sigma) \right]^{1/2} \epsilon/\sigma \tag{7}$$

where σ_D and σ are the rms noise after and before digitization and ϵ is the step size. The function $g_i(x)$ is defined by

$$g_i(x) = \operatorname{erf}[0.5(i + 1)x] - \operatorname{erf}[0.5ix] \tag{8}$$

where erf is the error function.[8] The step size ϵ may also be expressed by

$$\epsilon = S/2^N \tag{9}$$

where S is the maximum linear video signal and N is the number of bits in the ADC. Using the definition for the signal-to-noise ratio (SNR)

$$\mathrm{SNR} \equiv S/\sigma \tag{10}$$

one can relate ϵ/σ to SNR by

$$\epsilon/\sigma = \mathrm{SNR}/2^N \tag{11}$$

The ratio σ_D/σ may be computed using Eq. (7) with $\epsilon/\sigma = 1$, 2, and 3. These results are given in Table 1.

TABLE 1. Variation of Digitized rms Noise (σ_D) as a Function of Step Size (ϵ)

ϵ/σ	1	2	3
σ_D/σ	1.03	1.17	1.50

It is seen that if ϵ is equal to the analog rms noise σ, digitization introduces only an additional 3% noise. One can also use Eq. (11) to compute the number of bits required for a given SNR. This is given by

$$N = \text{INT}\,[\ln(\text{SNR}/m)/\ln 2] + 1 \tag{12}$$

where $m = \epsilon/\sigma$ and INT [·] indicates integer truncation operation. Figure 2 shows the relation between the SNR and the number of bits required.

For an SNR of 500:1, one needs an 8-bit ADC for $\sigma_D/\sigma = 1.5$ (50% error) and a 9-bit ADC for a 3% error. For many dynamic clinical applications, the ADC should also be fast enough to digitize the images in real

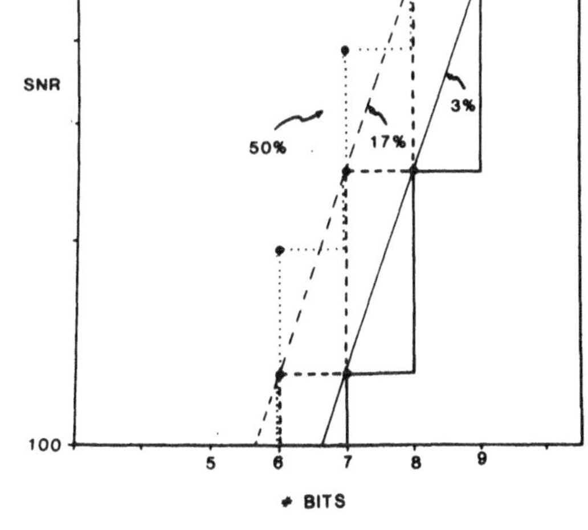

Figure 2. ln SNR versus number of bits.

time, i.e., 30 images/sec, with enough spatial resolution. A typical digitization format for real-time applications is 512 × 512 with 8 bits/pixel.

2.6. Image Acquisition Memory

After the data are digitized by the ADC, they are usually stored in a solid-state random access acquisition memory (RAM). For most radiologic applications, a matrix size of 512 × 512 pixels for the memory is sufficient. The size of the image matrix must be compatible with the TV camera resolution. Furthermore, each pixel must have at least as many bits as the number of bits of the ADC.

Larger memories are useful for image integration which is necessary in some applications. Even though acquisition memories as large as 2048 × 2048 pixels are also available, they do not seem to offer any additional advantage for quantitative applications.

2.7. Measurement of System Response

Here we will discuss the measurement of the TV camera and TV–II combination separately.

2.7.1. TV Camera Response

The purpose of this measurement is to determine the functional relationship given in Eq. (5) which relates the video signal V to the incident light level I. The TV camera response is measured by mounting the camera on an optical bench and varying the input light level incident on the camera.

The video voltage for each input light level I is measured to determine the response. The experimental results are given in Figure 3 for a 1-inch Syracuse Scientific plumbicon camera. In Figure 3 the video voltage versus the relative light input is plotted on a log–log graph. The response function for this camera was given by the simpler form shown in Eq. (6). If one substitutes Eq. (6) into Eq. (5) and takes the logarithm, one obtains

$$\ln V = \gamma \ln I + \ln a \tag{13}$$

The solid line in Figure 3 was obtained by a linear least-squares fit to the experimentally measured data (last point where camera saturation starts was not included). The fit had a correlation coefficient $r = 0.999$ and slope (γ) equal to 0.99. The maximum linear video signal from the camera was equal to 1 volt. The measured rms noise of the camera was determined

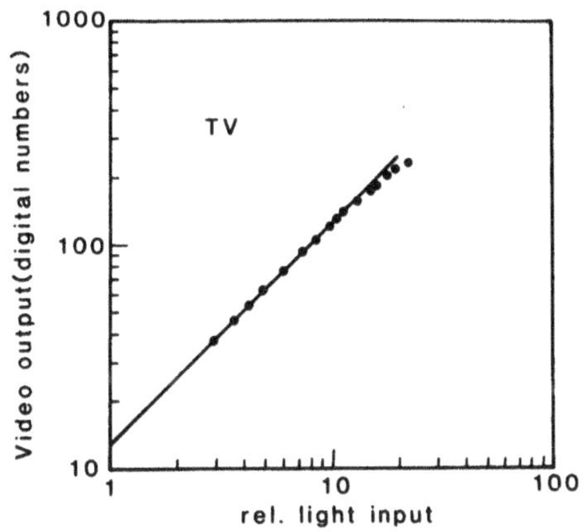

Figure 3. TV response as a function of relative light input.

to be 2 mV. Thus, the Syracuse Scientific camera used in our laboratory had a dynamic range of 500:1.

2.7.2. II–TV Response

Once the linear operating range of the TV camera is known, then the response of the II–TV system needs to be determined as a function of input X-ray exposure rate. This was done by varying the incident X-ray exposure rate by changing the X-ray mA while keeping the kVp constant. In this case, the video images were digitized and the II–TV response was determined from these measurements. The video voltage V is related to the input exposure rate X by

$$V = aX^{\gamma} \tag{14}$$

Taking the logarithm of both sides yields

$$\ln V = \gamma \ln X + \ln a \tag{15}$$

The experimental results for a Philips II–TV system are shown in Figure 4. The maximum linear voltage from the Philips camera was 920 mV and the rms noise was 4 mV. The Philips plumbicon camera had an SNR of 230:1. The straight line in Figure 4 is the II–TV response. The linear least-squares fit of Eq. (15) had a correlation coefficient $r = 0.99$ and $\gamma = 1.0$. Once the linear operation range of the TV camera and II–TV camera chain is known, one has to make sure to operate the system within this range so that there is a linear relationship between the output video

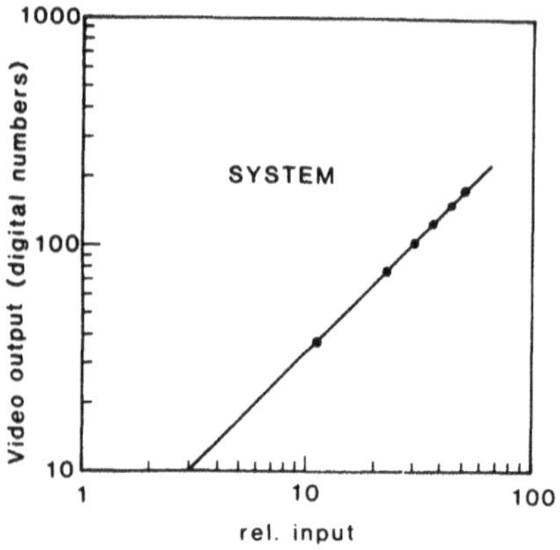

Figure 4. System response as a function of relative X-ray exposure.

voltage and the input X-ray exposure rate. Even if the response is not linear, one may still be able to linearize it by measuring the system response and using a lookup table.

2.8. System Spatial Resolution

Here we will closely follow the discussion given in Kruger *et al.*[9] If the focal spot of the X-ray tube has dimension b (mm), then the resolution limit (i.e., cutoff frequency) due to focal spot blurring is given by $f_{FS} = m/[b(m-1)]$ lp/mm as measured in the object plane for an object magnification m. For a 525-line TV camera, the vertical resolution is approximately 340 lines (see Hendee[10]). If we assume the same vertical and horizontal resolution for the TV camera, then the limiting spatial frequency is given by $f_c = 340/2D$ where D is the object size covering the full image field. In the object plane, this limiting resolution is given by $f_c = 340m/2D$. Finally, the limiting resolution in the object plane due to image digitization for a 512×512 image matrix is given by $f_D = 512m/2D$. In Figure 5, these three limiting frequencies are plotted as a function of magnification. It is seen that for $1 \leq m \leq 2$ (which is a reasonable range for most clinical applications), the bottleneck of spatial resolution is the 525-line TV camera. In this example, we assumed a 1.2-mm focal spot and a 9-inch II.

3. CHARACTERIZATION OF PHYSICAL DEGRADATION FACTORS

In this section we will present results which demonstrate effects of physical degradation factors on digital numbers measured with digital

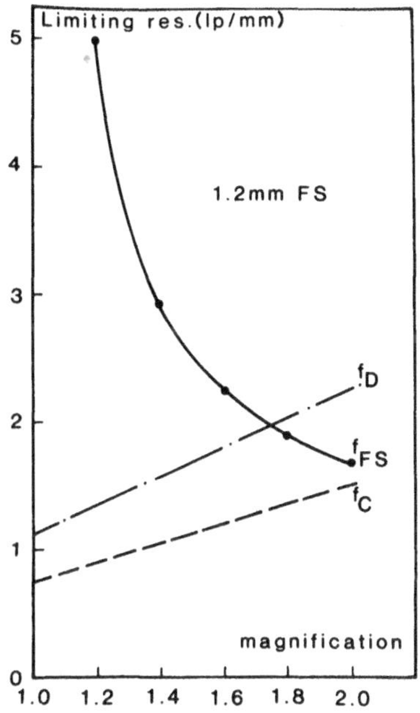

Figure 5. Resolution limits as a function of magnification. Solid curve, focal spot (1.2 mm); dashed curve, camera; dotted curve, digitizer cutoff frequencies for a 9-inch II.

fluoroscopic systems. The physical factors which will be considered are: beam hardening, X-ray scatter, and veiling glare. At this point, we will define a new function called the X-ray projection by

$$\Pr(L) = \ln[V(L = 0)/V(L)] \tag{16}$$

where $V(L)$ is the video signal (Eq. 5) at a location (x,y) for an object thickness of L. If the object is a composite material, then L is the upper limit of the integral in Eq. (3). For simplicity, we suppressed the dependence on (x,y) in Eq. (16) but it should be remembered that Pr is also a function of (x,y).

3.1. Beam Hardening

In order to study the effects of beam hardening alone, we have restricted the experimental setup to a narrow beam geometry, i.e., both the entrance and exit beams were collimated. Figure 6 illustrates the experimental setup. As a result of narrow beam geometry, we were able to eliminate scattered radiation within the object. The reduction of veiling glare[11–13] was accomplished by limiting the exposed field of view of the II to a small area (1 cm ∅). The reduction of the primary signal due to

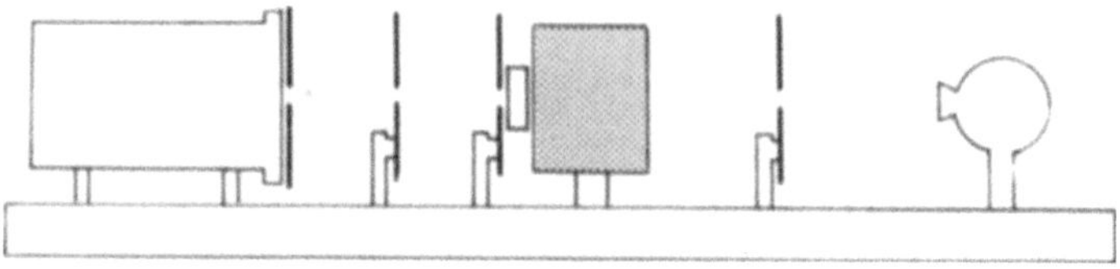

Figure 6. Narrow beam geometry on optical bench.

the limited II field of view was approximately 23% for the system used in these experiments. The reduction is due to the fact that light from the center escapes toward the edges, but since the remainder of the II was covered with lead, no light travels from the periphery to the center to compensate for the loss of light from the center to the periphery.

With this setup, Eqs. (2)–(6) can be combined to yield

$$V_1(x,y) = \beta \int D(x,y:E)P(x,y:E)dE \tag{17}$$

where the transfer function was taken to be $T(x,y;x',y') = \delta(x - x')\delta(y - y')$ and $\beta = a\alpha$. An experimental investigation of Eq. (17) was done by using various thicknesses of lucite as a scatter phantom, along with changes in X-ray tube filtration and iodine thickness in the X-ray beam. The imaging chain consisted of a Philips dual-mode (9 inch/5 inch) II, Model 9807 201 02 701. The X-ray tube was a Dunlee DU-175 0.3/1.0 dual-mode focal spot type. The X-ray generator was a 3-phase, 12-pulse system manufactured by Picker. The TV camera was a 1-inch Philips plumbicon system. Video images were digitized and stored in the memory of an American Edwards Laboratories Cardiac 1000 video image processor. The images were digitized into 512 × 512 format and 8 bits/pixel. A square ROI was chosen and the mean and standard deviation within the region were computed. The mean value corresponds to the digital value of the video signal due to transmitted X-rays. In this measurement, the video signals were not logarithmically amplified prior to digitization.

The experiment was designed to study the amount of beam hardening due to iodine in the presence of a fixed thickness of lucite. We varied the concentration of Renografin-76 which was placed behind the lucite object. The experiment was repeated twice, using 0- and 4-mm added aluminum filtration at 70 kVp and 20 mA. The measured projection values, as defined in Eq. (16), are plotted in Figure 7. For a monoenergetic X-ray beam, i.e., $D(x,y:E) = D\delta(E - E_0)$ where E_0 is the energy, the projection given in Eq. (16) is equal to

$$\mathrm{Pr}_{\mathrm{mono}}(L) = \mu(E_0)L \tag{18}$$

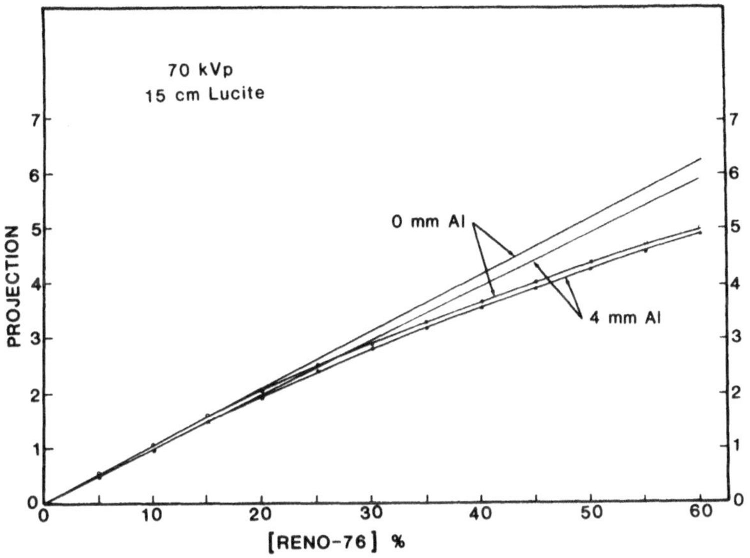

Figure 7. X-ray projection for the primary beam as a function of Renografin-76 concentration.

where $\mu(E_0)$ is the linear attenuation coefficient of a homogeneous material at an energy E_0 and thickness L (in this example, concentration). Thus, for a monoenergetic X-ray beam, the projection is a linear function of thickness. Nonlinear response in this case is due to beam hardening since the imaging system was shown previously to be linear. The straight lines in Figure 7 were obtained using the linear least-squares fitting technique with the experimental data within the low concentration range (0–5%). They also represent the monoenergetic extrapolation to the data. The deviation of the experimental points from the straight line is an indication of beam hardening with this system using iodine in the presence of 15-cm lucite. As seen from Figure 7 and as expected, the deviation is larger than 0-mm added aluminum filtration.

3.2. X-Ray Scatter

An experimental study of the effect of object scatter without the influence of veiling glare was undertaken in the following manner. A lead cover with a small circular aperture (1 cm ∅) at the center was used to cover the II face as described in Section 3.1. The lead collimators in front and back of the scatter medium were removed. The experimental setup is illustrated in Figure 8. The distance between the exit face of the lucite

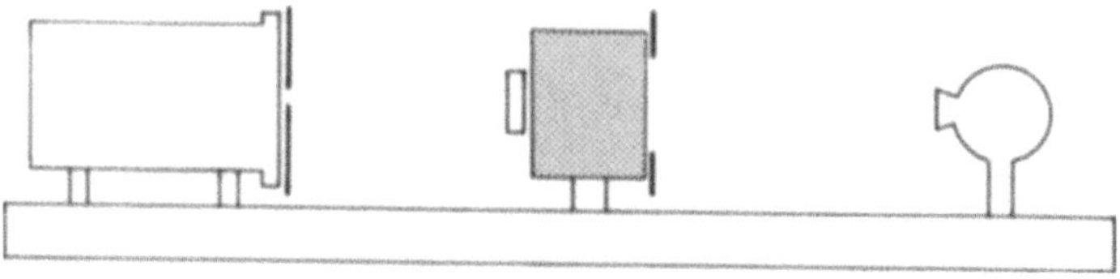

Figure 8. Open beam geometry for measuring X-ray scatter.

and II input was 25 cm. A 15-cm-diameter circular collimator was placed on the entrance face of the lucite which geometrically restricted the transmitted primary X-rays to just within the diameter of the intensifier input (9-inch diameter). With this setup, the detected video signal within the circular aperture was given by

$$V_2(x,y) = c \int D(x,y{:}E)N(x,y{:}E)dE \qquad (19)$$

where N is given by Eq. (2) and includes both the primary and scattered photons detected at location (x,y). Due to removal of veiling glare with this experimental setup, we substituted $T(x,y;x',y') = \delta(x - x')\delta(y - y')$ in deriving Eq. (18) from Eq. (4). The amount of beam hardening was determined in Section 3.1 and can be corrected for, so as to eliminate beam-hardening artifacts and study the effect of X-ray scatter alone.

An experimental study of Eq. (18) was done using 15-cm lucite with various concentrations of 2-cm-thick Renografin-76 solution as discussed in Section 3.1. The video signal was measured both with and without an X-ray grid on the II. The measured signal, primary plus scatter, is denoted by P + S in Figure 9. The primary signal (P) with and without the grid was measured using the experimental setup shown in Figure 6. The scatter component (S) was computed by taking the difference of (P + S) and (P) and is also shown in Figure 9. It is seen that for large concentrations of iodine, the (P + S) curves approach a constant value asymptotically. This is due to the fact that for large concentrations of iodine, neither the primary nor the scattered photons pass through iodine. The X-rays which reach the open aperture in this case are the ones which are scattered within lucite without passing through the iodine. Thus, this scattered component is independent of the iodine concentration and hence the constant value. The total number of scattered photons reaching the open aperture on the II (see Figure 8) can be expressed as

$$S_{\text{total}} = S_{\text{iodine}} + S_{\text{outside}} \qquad (20)$$

where S_{iodine} and S_{outside} are the number of scatterd photons passing through iodine and the outside region, respectively. In Figure 10, S_{iodine} was com-

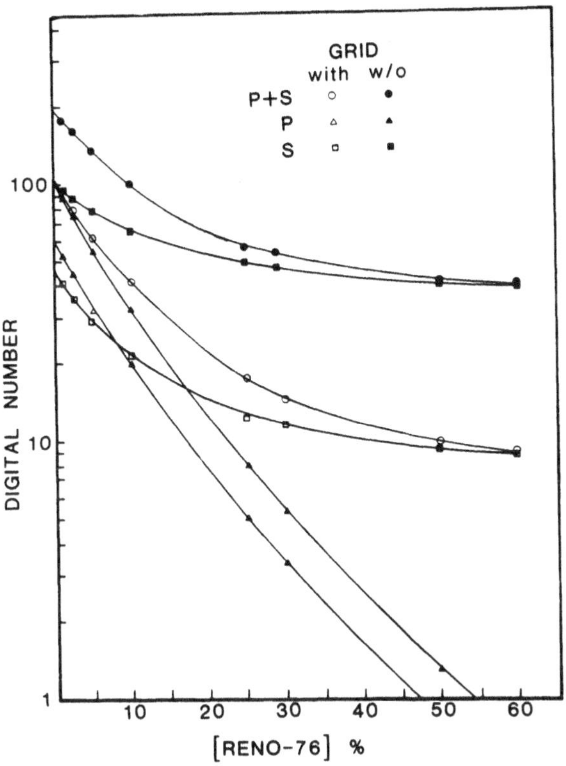

Figure 9. Measured digital numbers versus Renografin-76 concentration. P, primary; S, scatter; P + S, primary and scatter.

puted by subtracting the constant asymptotic value (i.e., for large iodine concentrations) from S_{total}. It is seen from Figure 10 that when S_{iodine} versus concentration is plotted semilogarithmically, one obtains the expected straight line relationship. The situation is similar with or without the X-ray grid. The measurements presented in Figures 9 and 10 were done with 4-mm Al added filtration at 70 kVp. Figure 11 shows the measured projection (see Eq. 16) as a function of iodine concentration. In the same figure, we plotted the polychromatic primary projection taken from Section 3.1 as well as the monoenergetic extrapolation. We see that the polychromatic projection deviates from the straight line monoenergetic one as the concentration of iodine is increased. The deviation from linearity is more substantial when the scatter is included. As one would expect, the situation is slightly better with than without the grid. The results presented in this section included the effects of scatter and beam hardening on the measured video signal and deviation from linearity of the projection.

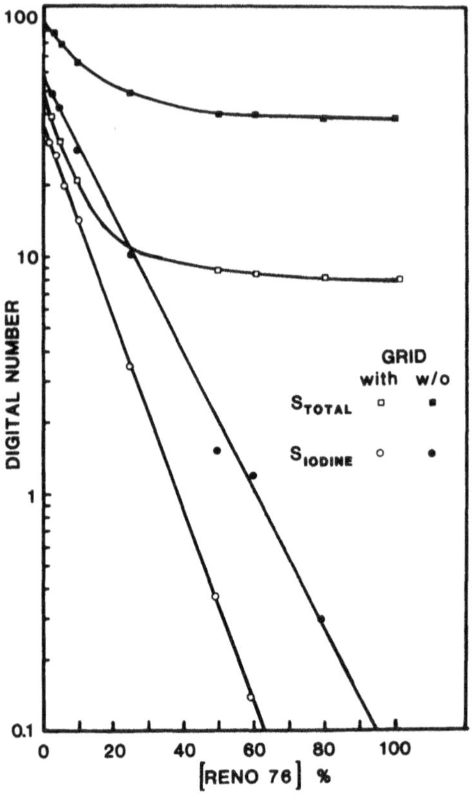

Figure 10. Computed scatter contribution versus iodine concentration.

3.3. Veiling Glare

The final study included all the physical degradation factors: beam hardening, X-ray scatter, and veiling glare. In this case, the detected video signal at a location (x,y) is given by Eq. (4). An experimental realization of this was achieved by removing the lead cover on the II, thus exposing the whole input phosphor during the experiments. The setup is illustrated in Figure 12. The experiment was performed by changing the concentration of Renografin-76 within a 2-cm-thick bottle in the presence of 15-cm-thick lucite. The images were acquired with 70 kVp at 20 mA. The experiment was repeated both with and without an X-ray grid. The mean digital number within a small square region at the center of the image was measured for all cases. These numbers, indicated by (P + S + VG), are shown with and without a grid in Figure 13. Also shown are the primary measurements performed by using the setup described in Section 3.1. The two curves for the primary with and without the grid are separated by a

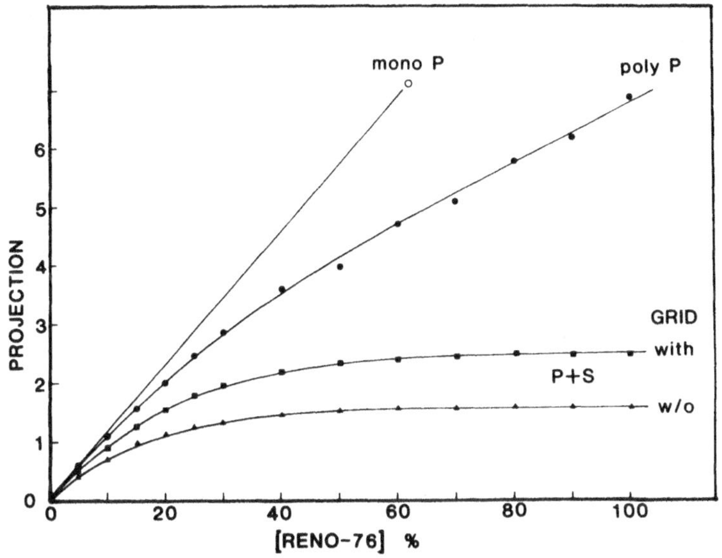

Figure 11. Projection for various cases as a function of iodine concentration.

constant amount due to constant grid transmission. The measured digital number for (P + S + VG) approaches a constant value as the concentration of Renografin-76 is increased in a manner similar to that of Section 3.2. Using the data presented in Figure 13, we computed the projection for this case. The results are shown in Figure 14. The projection for (P + S) and (P + S + VG) both become constant as the iodine concentration becomes large. Including veiling glare makes the situation a little worse. As one might expect, the use of an X-ray grid somewhat improves the overall response. In the same figure, we show the primary projection which is degraded only by beam hardening. Finally, we show the projection data with a grid for all three cases in Figure 15, and include the linear extrapolation for each case, indicated by "mono." These straight lines

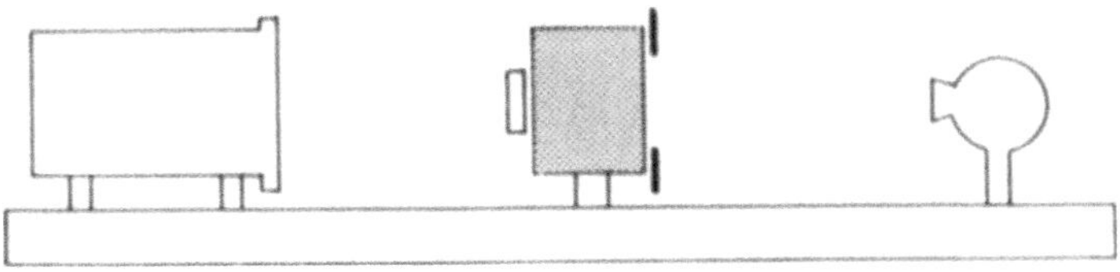

Figure 12. Open field geometry to include veiling glare, scatter, and beam hardening.

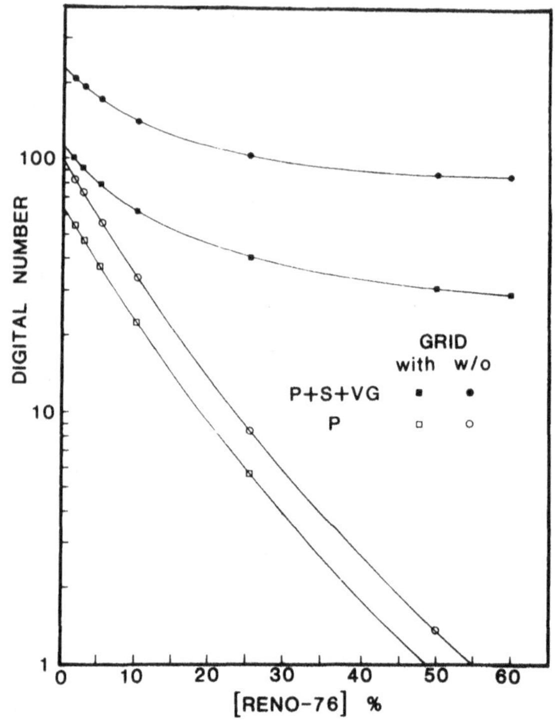

Figure 13. Measured digital numbers versus iodine concentration. P, primary; P + S + VG, all degradations included.

were obtained by fitting a straight line through the first two or three experimental points. A substantial deviation from linearity exists for all three cases. It is the nonlinearity projection data which cause errors in measurements. These errors are the subject of the next section.

4. EFFECT OF DEGRADATION FACTORS ON VIDEODENSITOMETRIC VOLUME MEASUREMENTS

An experimental investigation of the nonlinear behavior of X-ray projection as a function of object thickness was presented in Section 3. In these studies, we used a 2-cm-thick bottle filled with various concentrations of iodine and placed it behind a 15-cm-thick lucite block. By varying the experimental setup, we were able to measure the projection as a function of iodine concentration under various experimental conditions. The projection data presented in Figures 14 and 15 were for a single

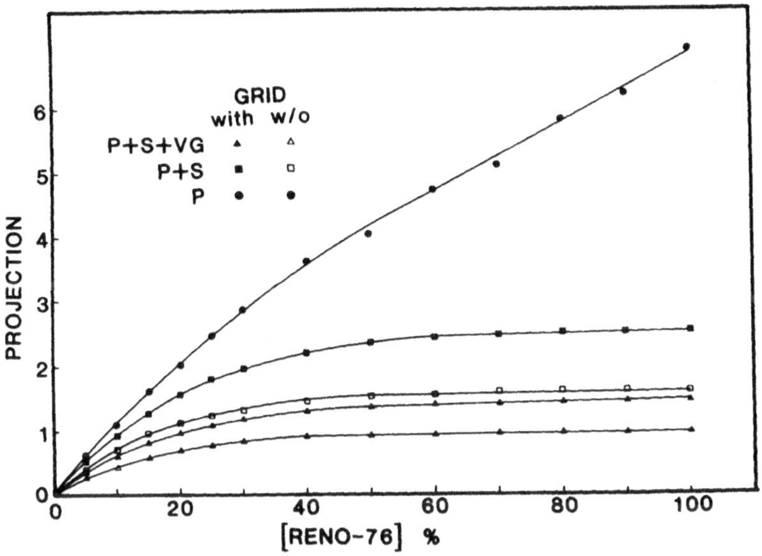

Figure 14. Projection for P, P + S, and P + S + VG as a function of iodine concentration.

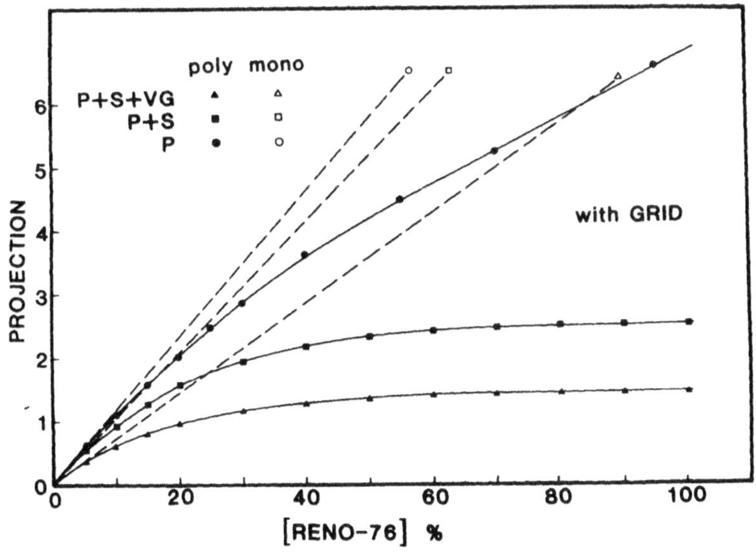

Figure 15. Projection using an X-ray grid for P, P + S, and P + S + VG (solid curves). Dashed lines are respective monochromatic extrapolations.

ray passing through the object. The experimental projection data were used in a mathematical simulation of a three-dimensional object filled with Renografin-76 under the same experimental conditions. The object chosen was an ellipsoid of revolution. The two minor axes were taken to be equal while the major-to-minor-axis ratio was assumed to be 2:1. In the first simulation, the major axis was taken to be along the X-ray beam direction (a parallel beam geometry was also assumed). This situation is clinically similar to left ventricular imaging using the left anterior oblique (LAO) view with X-rays traveling along the long axis of the ventricle. In the second situation, we assumed the X-ray beam to be perpendicular to the long axis of the ellipsoid of revolution. This view clinically represents the right anterior oblique (RAO) imaging of the left ventricle. From here on, we will call these two different simulations LAO and RAO, respectively. The simulated volume of the ellipsoid was varied from 2 to 200 cm^3 while keeping the ratio of the axes constant. All the different size ellipsoids were assumed to be filled with a 10% concentration of Renografin-76. In order to compute the gray levels within the simulated images, we first computed the cord length through the ellipsoid corresponding to a given pixel location within the image, and then assigned the experimental X-ray projection corresponding to this specific cord length to that pixel. Since the cord lengths vary continuously from one point to another within the object, the simulation requires a knowledge of the X-ray projection as a function of object thickness continuously. This was accomplished by fitting a polynomial to the experimental projection data shown in Figures 14 and 15. The specific function used is given by

$$\Pr(x) = \sum_{i=1}^{M} a_i(x)^i \tag{21}$$

where M is the order of the polynomial, x is the thickness of 10% Renografin-76, and a_i are the coefficients obtained by a least-squares fitting of Eq. (21) to the experimental data. In all studies, it was sufficient to choose $M = 4$. Two examples of the simulated ellipsoid images are shown in Figure 16. In Figure 16a is shown a simulated RAO image of the ellipsoid of revolution. The LAO image, which is in a direction perpendicular to the circular cross section, is shown in Figure 16b.

4.1. Absolute Volume Measurements

By using the experimental projection data degraded by beam hardening, scatter, and veiling glare, we studied the order of magnitude of the

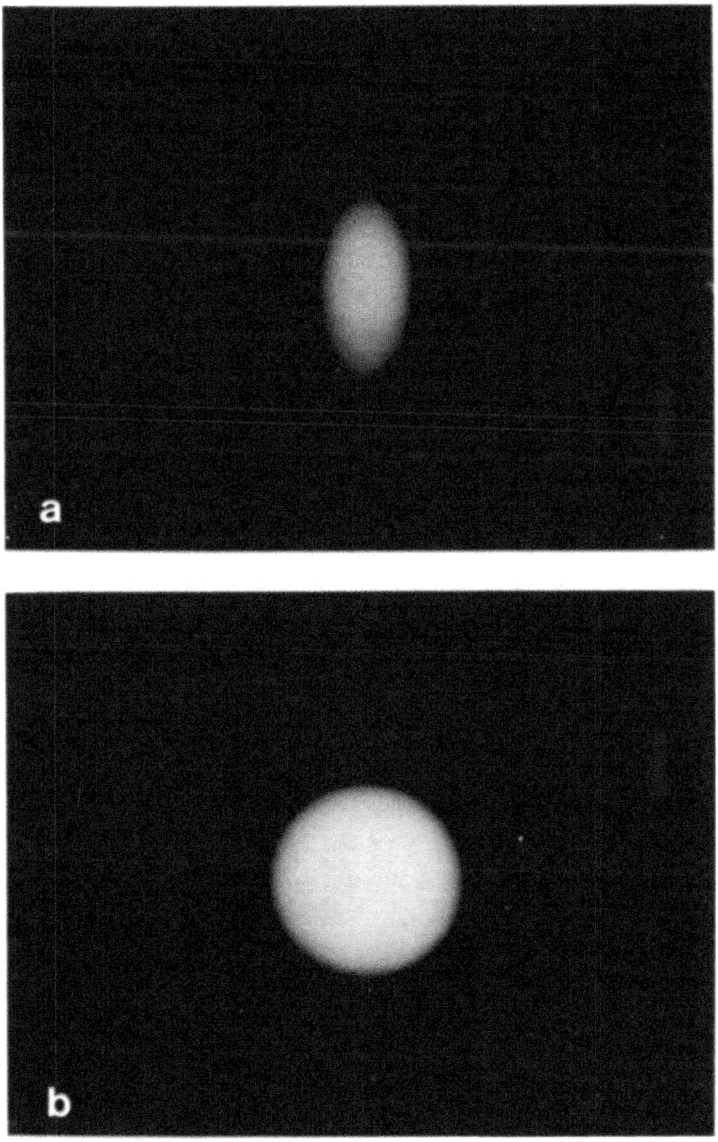

Figure 16. (a) Simulated RAO-view ellipsoid of 50-cm^3 volume (P + S with grid). (b) Simulated LAO-view ellipsoid of 50-cm^3 volume with different magnification (P + S with grid).

errors introduced by various physical degradation factors. Once a simulated image was generated, the total videodensitometric signal was measured by integrating the gray levels within the object. Figure 17 illustrates the selection of a region-of-interest (ROI) around the object. The integrated net signal within the ROI is supposed to be proportional to the volume of the object. We are now in a position to verify the correctness of this statement. In Figure 18, we show the videodensitometrically measured volume versus the actual ellipsoid volume. The monochromatic curve was obtained by fitting a straight line through the first few experimental points. This curve was also used to normalize the measured volume so that it is given in units of cm^3. As a result of the normalization, P_0 represents the line of identity. In the same figure, we also show the curves for the primary beam and the realistic case when all the degradation factors were included. The measured volumes for the RAO view using the primary beam alone are slightly underestimated, maximum being at 200 cm^3. This is due to additional beam hardening by the iodine within the ellipsoid. In the RAO view, the largest thickness is 5.8 cm for the 200-cm^3 volume (i.e., minor axis). When beam hardening, scatter, and veiling glare are all included, the deviation from the line of identity becomes larger. Notice that, in spite of this large departure from the line of identity, the curve is still fairly linear. The meaning of this observation

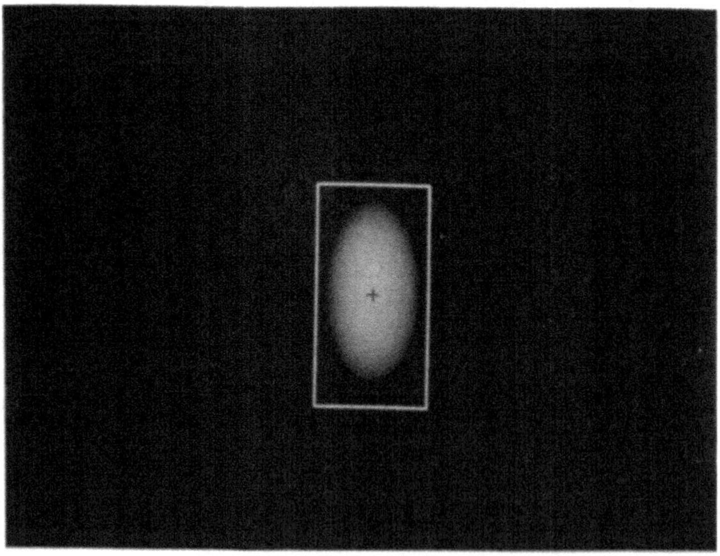

Figure 17. Selection of an ROI around the object.

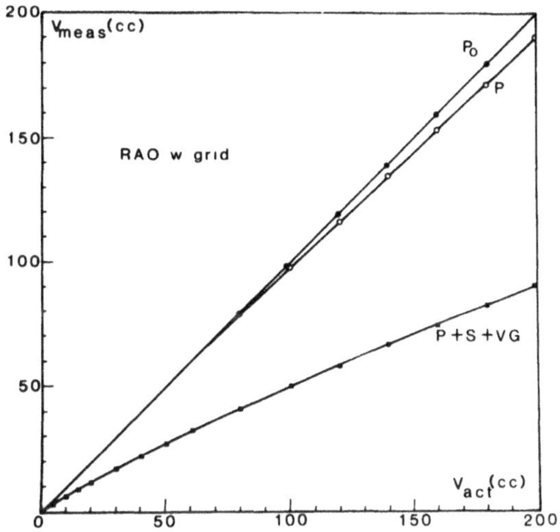

Figure 18. Measured volume versus actual one for the RAO view for various cases.

is that, while the absolute value measurements may be underestimated greatly, relative volume measurements such as ejection fraction would be underestimated less. Figure 19, representing the LAO view, shows what happens when we image the ellipsoid along the long axis. The notation is the same as for Figure 18. The underestimation of volumes obtained using the primary beam are larger than the previous (RAO) case. The largest thickness traversed in the LAO case is the major axis, 13 cm for

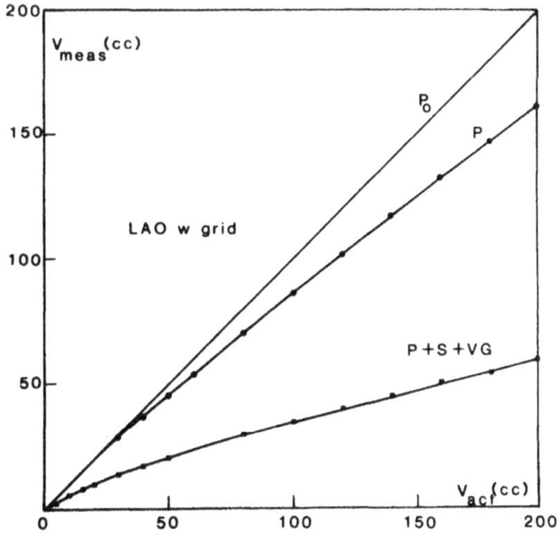

Figure 19. Same as Figure 18 but for LAO view.

the 200-cm^3 volume. Thus, we have a larger amount of beam hardening due to added iodine, and, hence, the larger deviation from the line of identity. If we study the curve obtained by using all the degradation factors, we see that it is also underestimated more as compared to the RAO case (Figure 18). However, a close examination of the curve indicates that most of this deviation is due to a larger amount of beam hardening. In other words, the separation between the primary and the full curves in Figure 19 is almost identical to that of Figure 18.

4.2. Relative Volume Measurements

In this subsection, we will define a quantity which is similar to the ejection fraction used in cardiology. It is given by

$$EF = 100(1 - V_{<}/V_{>}) \tag{22}$$

where $V_{<}$ and $V_{>}$ correspond to end-systolic and end-diastolic volumes, respectively. We will also define a percentage error (ϵ) by

$$\epsilon = 100[1 - EF(measured)/EF(actual)] \tag{23}$$

In Figure 20, we show the percentage error in EF for two different end-diastolic volumes, i.e., for $V_{>} = 80$ and 200 cm^3. These curves were obtained for the case when all the degradation factors were included. In the RAO view, the largest percentage error occurs for large end-diastolic volumes with small ejection fractions. Intermediate, end-diastolic error

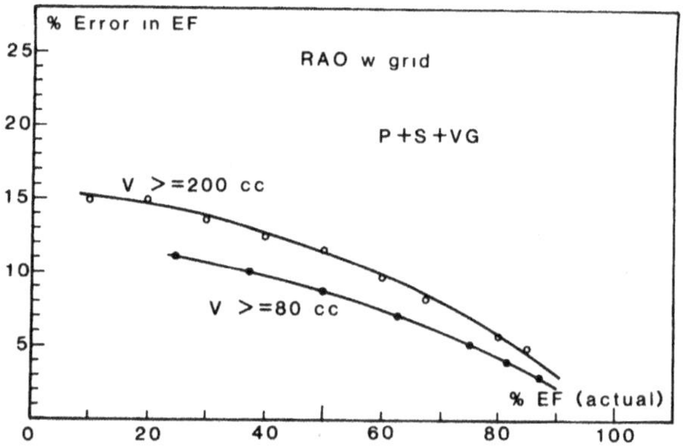

Figure 20. Percentage error in EF as a function of actual EF and various end-diastolic volumes—RAO view.

curves lie in between the two shown in Figure 20. It is seen from Figure 20 that for an ejection fraction of 70% and an end-diastolic volume of 200 cm^3, the error is 8%. If the actual ejection fraction is 30% for the same end-diastolic volume, the error goes up to 14%. The results shown in Figure 20 were obtained for the RAO view. Similar results for the LAO view are shown in Figure 21. Here, the errors are slightly larger compared to the RAO case.

5. TECHNIQUES FOR REDUCTION OF DEGRADATION FACTORS

In this section, we will discuss and review some techniques for improving the accuracy of measurements performed with digital fluoroscopic systems. The degradation factors which will be considered are the ones mentioned in the previous sections, namely beam hardening, X-ray scatter, and veiling glare.

5.1. Veiling Glare

The equation which describes the relationship between the incident X-ray flux and the output video signal is given by

$$V(x,y) = \beta \iiint D(x',y':E)T(x,y;x',y')N(x',y':E)dx'dy'dE \tag{24}$$

where we assumed that a plumbicon TV camera with $\gamma = 1$ is used. Using the Seibert model,[14] we will expand T in terms of a function h, which is the stationary point spread function (PSF),

$$T(x,y;x',y') \simeq h(x - x',y - y') + \Delta h \tag{25}$$

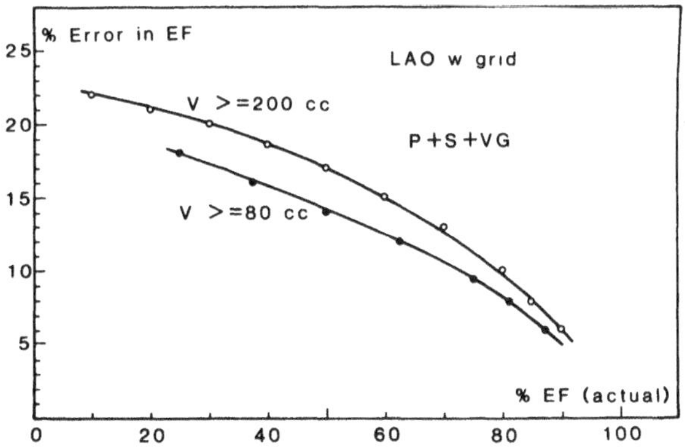

Figure 21. Same as Figure 20 but for LAO view.

where Δh represents higher-order nonstationary terms. We have shown earlier that Δh is negligible compared to h.[14] In other words, PSF depends only on the difference of and not on each variable separately. This allows rewriting of Eq. (25) as a convolution relationship which is given by

$$V(x,y) = [h**U]_{(x,y)} \tag{26}$$

where $**$ is a two-dimensional convolution operation and U is the image which is not degraded by the TV–II system. It is defined by

$$U(x,y) = \beta \int N(x,y:E)D(x,y:E)dE \tag{27}$$

A derivation of the veiling glare PSF was outlined in a previous publication.[15] This derivation was based on the assumption of linear system response, circular symmetry, and spatial invariance of the II PSF. It was shown that the theoretical PSF can explain the experimental results to a high degree of accuracy. The analytical form of PSF is given[12] as

$$h(r) = (1 - \rho)\, \delta(r)/r + \rho \exp(-r/k)/(2kr) \tag{28}$$

where ρ and k are the parameters describing the fraction of strongly scattered light and its mean propagation distance, respectively. The symbol δ in the above equation is the Dirac delta function. The relationship between the undegraded image, $U(r,\theta)$, and the actual detected image, $V(r,\theta)$, in polar coordinates is given by

$$V(r,\theta) = [h**U]_{(r,\theta)} \tag{29}$$

Equation (29) mathematically represents the two-dimensional (2D) convolution of the II PSF $[h]$ with the undegraded image $[U]$ resulting in the detected image $[V]$. In this equation, $**$ represents the 2D convolution process, and (r,θ) the polar coordinate system chosen with which to describe the process. The PSF $[h]$ is represented as a linear and rotationally symmetric function. Thus, we neglect the higher-order terms expressing the nonlinear and nonstationary system response based on our previous findings.[15]

The 2D convolution in spatial domain given in Eq. (29) can be represented in the frequency domain by

$$F_2[V] = F_2[U]\, F_2[h] \tag{30}$$

where $F_2[\cdot]$ is a 2D FT operation. The right-hand side of Eq. (30) is a multiplication of the frequency domain representation of the undegraded image with the PSF on a point-by-point basis. Inverting $F_2[h]$ in Eq. (30) and taking the inverse FT yields

$$U(r) = V(r) ** F_2^{-1} \tag{31}$$

where $F_2^{-1}[\cdot]$ is the inverse 2D FT operation. The undegraded image U can be obtained by direct 2D deconvolution on the acquired spatial domain image with an inverse filter[16] if the PSF $[h]$ is known.

The spatial extent of the veiling glare PSF is considerable, as indicated by the magnitude of k. The mean value is typically 2–3 cm, while the pixel size has millimeter dimensions. In order to properly deconvolve the PSF, a large number of pixels becomes mandatory to fully describe the inverse filter. The long range of the filter prohibits the implementation of Eq. (31) in the spatial domain due to the excessive computation time required. Instead, the efficiency and speed of the fast Fourier transform[17] permits rapid deconvolution to be carried out in the frequency domain by solving for $F_2[U]$ in Eq. (30) and inverse transforming the result:

$$U = F_2^{-1}[F_2[V]\,\{1/F_2(h)\}] \tag{32}$$

Since we know the analytical form of h, $1/F_2[h]$ can also be analytically derived, with the tremendous benefit of avoiding singularities and oscillatory behavior present in a digital calculation caused by finite sampling width, truncation errors, and problems associated with noise.

The veiling glare PSF can be transformed into the frequency domain by Fourier methods:

$$\hat{H}(f) = \int_0^\infty \int_0^{2\pi} h(r)\, e^{-2\pi i r \cdot f}\, r dr d\theta \tag{33}$$

In this equation, f is the spatial frequency (cm^{-1}) and $\hat{H}$ is the frequency domain representation of h. Taking advantage of the circular symmetry, the Fourier transform becomes a Hankel transform, giving

$$\hat{H}(f) = 2\pi \int_0^\infty h(r) J_0(2\pi r f) r dr$$

and substituting Eq. (28) results in

$$\hat{H}(f) = 2\left[\frac{\rho}{2k}\int_0^\infty e^{-r/k} J_0(2\pi r f) dr + (1\text{-}\rho)\int_0^\infty \delta(r) J_0(2\pi r f) dr\right] \tag{34}$$

where J_0 is the zero-order Bessel function and δ is the Dirac function.

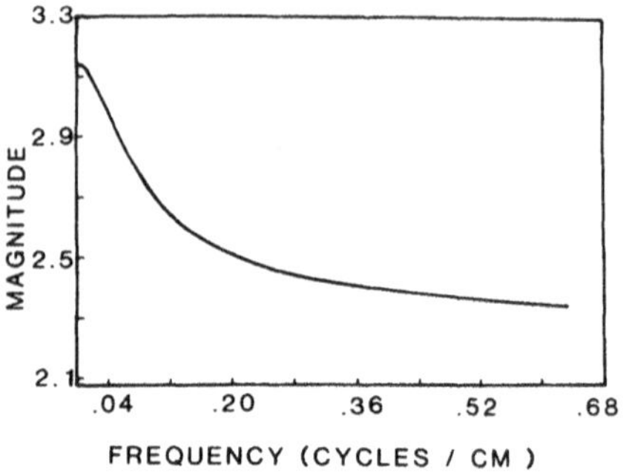

Figure 22. Veiling glare PSF in the frequency domain.

Integrating and solving Eq. (34) results in the frequency domain PSF

$$\hat{H}(f) = \Pi \left[\frac{\rho}{\sqrt{1 + (2\pi k f)^2}} + (1 - \rho) \right] \tag{35}$$

The form is illustrated in Figure 22. Only a profile along a radius is shown. The complete filter may be obtained by rotating about the z axis.

Inversion of Eq. (35) gives the solution of the inverse frequency filter

$$\hat{H}^{-1}(f) = \frac{1}{\pi} \left[\frac{\sqrt{1 + (2\pi k f)^2}}{\rho + (1 - \rho)\sqrt{1 + (2\pi k f)^2}} \right] \tag{36}$$

Its form is plotted in Figure 23. As illustrated, the filter is essentially high pass in nature; it retains higher frequencies, and selectively attenuates lower frequencies which contribute to the veiling glare degradation in the image.

Multiplication of this filter on a point-by-point basis with the frequency domain degraded image yields a product in the frequency domain

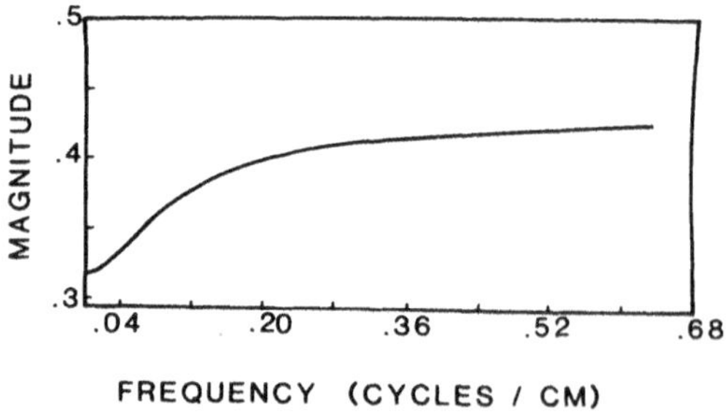

Figure 23. Inverse filter in the frequency domain.

that has the lower frequencies deamplified. Application of an inverse 2D FT with appropriate scaling factors results in an approximate estimate of the original undegraded image. The spatial domain representation of the inverse filter with values of $\rho = 0.28$ and $k = 2.7$ cm is illustrated in Figure 24. The graph indicates the negative contribution of values to the point being deconvolved, counteracting the spread of light at adjacent pixels. The spatial extent of the filter is extensive, with significant negative weighting at large distances from the center pixel.

5.1.1. Deconvolution of Lead Disk Images

A lead disk of 1/8-inch thickness was placed on the II input phosphor cover (no grid) and a digital fluoroscopic image was acquired. The focal spot to input phosphor distance was 1.5 m. No scattering medium was present. The image is shown in Figure 25a. Notice that even though the lead disk was thick enough to stop any X-ray transmission through it, the

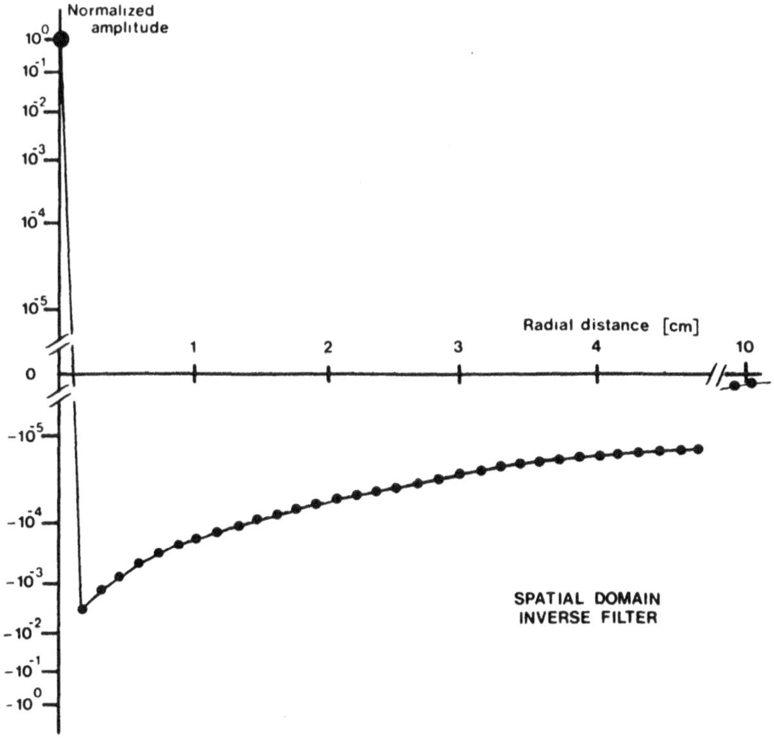

Figure 24. Normalized inverse filter in the spatial domain.

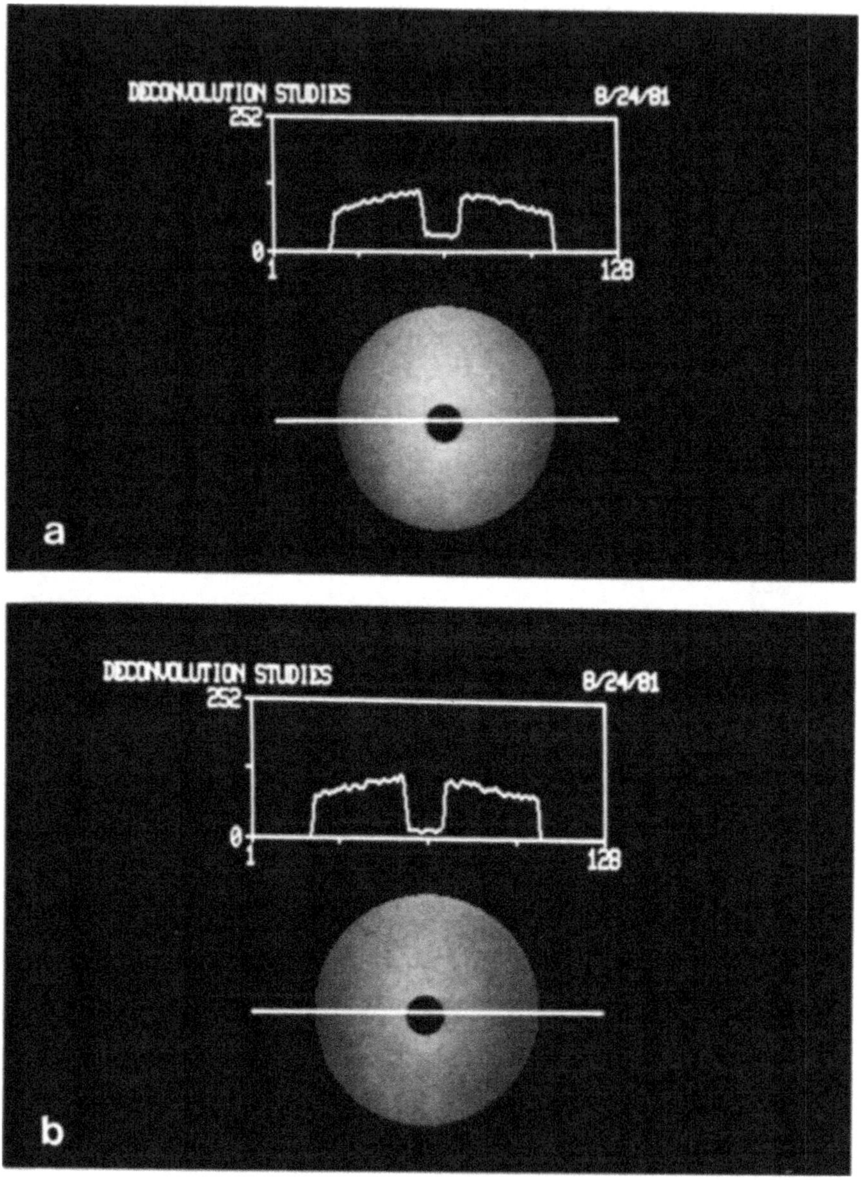

Figure 25. (a) A lead disk image and a diametric profile without deconvolution. (b) Same image after glare deconvolution.

video profile displayed in Figure 25a indicates a nonzero reading under the lead disk. This is due to the glare effect under discussion all along. The light produced within the open areas leads into the region which is covered by the disk. After applying the glare deconvolution technique discussed in Section 5.1, we obtained the deconvoluted image shown in Figure 25b. The signal under the lead disk is close to zero. If the deconvolution process were to be totally accurate, we could have obtained a zero reading under the disk.

5.1.2. Effects of Glare Deconvolution on Volume Measurements

In Section 5.1.1, we studied the effect of veiling glare deconvolution in the absence of any scatter medium. We will now discuss what happens when a scatter medium is present. Furthermore, we will also investigate what happens to volume measurements. We used the images of the iodine bottle filled with different concentrations of Renografin-76 and placed behind a 15-cm-thick lucite phantom. As mentioned earlier, all three of the physical degradation factors we have been concerned with were present in these images. We deconvolved each image (containing different concentrations of iodine) with the appropriate veiling glare filter and computed the X-ray projection again. These results are shown in Figure 26. Here, the projections for the primary beam, primary plus scatter, and primary plus scatter plus veiling glare are shown. For comparison, we

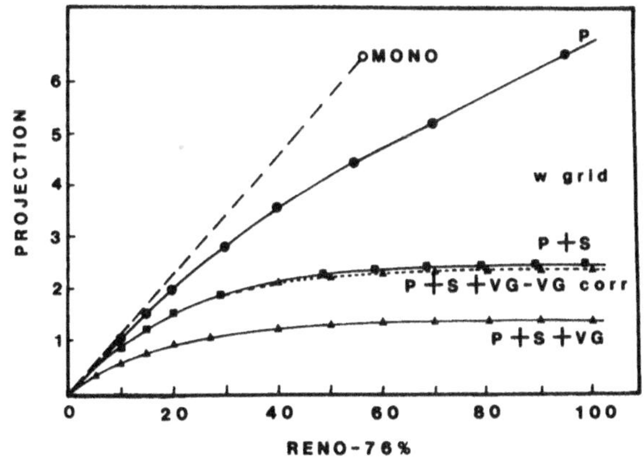

Figure 26. Summary of projection versus concentration of Renografin-76 for primary (P), primary plus scatter (P + S), and primary plus scatter plus veiling glare (P + S + VG) polychromatic projections and their monoenergetic extrapolations. Also shown are the deconvolved projection data using an inverse PSF: ρ = 0.241, k = 2.97 cm as measured for the II used in the experiments (small dashed curve, veiling glare corrected).

also indicate the projection obtained after deconvolution of the veiling glare PSF. The deconvolution process brings the (P + S + VG) projection curve almost on top of the (P + S) one. For perfect deconvolution, (P + S) and "VG corrected" curves should overlap. It is seen from Figure 26 that the projection obtained after glare deconvolution does not contain much glare effect.

We used deconvolved projection data and studied the effect of the glare reduction technique on videodensitometric absolute volume measurements. In this investigation the "LAO" ellipsoid of revolution phantom discussed in Section 4.1 was used. The results are shown in Figure 27. It is seen that the experimentally obtained (P + S) curve and the mathematically obtained one, i.e., deconvolved (P + S + VG) curve, almost overlap. Thus, we conclude that the deconvolution technique discussed in Section 5.1 is quite good in reducing the errors introduced by veiling glare.

5.2. X-Ray Scatter

We saw in the last section that after the deconvolution of veiling glare, we are left with the system undegraded image U which was defined by Eq. (27). The function N used in Eq. (27) was defined in Eq. (2) as

$$N(x,y:E) = P(x,y:E) + S(x,y:E) \tag{2}$$

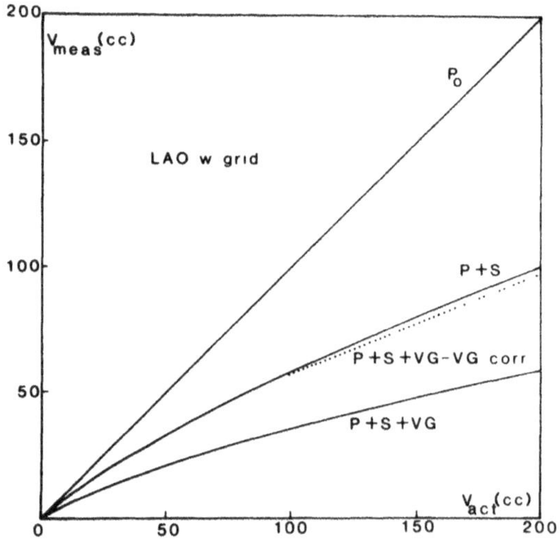

Figure 27. Measured volume versus actual one for the cases shown in Figure 26.

where P and S are the primary and scattered photons incident on the II, respectively.

As we have seen in Section 3.2 (Figure 10), some fraction of the scattered photon distribution may appear as a low-frequency version of the primary photon distribution, i.e., carries some information about the spatial distribution of attenuators within the object. Mathematically, we can express this as

$$N(x,y:E) = \int P(x',y':E)h_s(x - x',y - y':E)dx'dy' \tag{37}$$

where the scatter PSF is defined by

$$h_s(x,y:E) = \delta(x)\delta(y) + \left[\frac{s(x,y:\epsilon)}{p(x,y:\epsilon)}\right] g(x,y:\epsilon) \tag{38}$$

In Eq. (38), $\delta(\cdot)$ is a one-dimensional delta function and $g(x,y:E)$ is a function to be determined later. We also notice that the quantity in brackets is the scatter-to-primary ratio (SPR). We will now approximate the SPR by

$$\text{SPR} \simeq S(0,0:\bar{E})/P(0,0:\bar{E}) \tag{39}$$

where $\bar{E}$ is the effective beam energy for the X-ray beam. We can now rewrite Eq. (38) as

$$h_s(x,y:E) = \tilde{h}_s(x,y:\bar{E}) + \Delta h_s \tag{40}$$

where

$$\tilde{h}_s(x,y:\bar{E}) = \delta(x)\delta(y) + \text{SPR} \cdot g(x,y:\bar{E}) \tag{41}$$

We also substituted $\bar{E}$ for E in g. As a first-order approximation, we will now assume that Δh_s in Eq. (40) is small compared to $\tilde{h}_s$. Thus, we can rewrite Eq. (38) as

$$N(x,y:E) \simeq \int P(x',y':E)h_s(x\text{-}x',y\text{-}y':\bar{E})dx'dy' \tag{42}$$

The last term in Eq. (41) is similar to the scatter PSF as defined by Barrett and Swindell.[18] They have shown that in the single-scatter approximation,

$$g_{\text{single}}(r:\bar{E}) = \exp[-\alpha(\bar{E})r^2] \tag{43}$$

where α is an energy-dependent coefficient. If one takes multiple scattering into account, Eq. (43) can be generalized into

$$g_{\text{multi}}(r{:}\bar{E}) = \exp[-\alpha(\bar{E})r^2] \sum_{i=1}^{M} \beta_i(\bar{E})(r)^{2i} \tag{44}$$

where M is the number of terms in the expansion and $\beta_i(\bar{E})$ are energy-dependent coefficients. Equations (43) and (44) indicate that for a thin object ($M = 0$), the scatter PSF is a Gaussian. For thicker objects, multiple scattering becomes significant and one has to use the more general formula given in Eq. (44). The higher-order terms make the Gaussian function broader, indicating a less structured two-dimensional scattered photon distribution. This, of course, is intuitively correct since the directional information the scattered photons carry is lost as a result of multiple scattering. If we substitute Eq. (42) in Eq. (27), we obtain

$$U(x,y) = (P{*}{*}\tilde{h}_s)_{(x,y)} \tag{45}$$

where

$$\tilde{P}(x,y) = \int P(x,y{:}E)D(x,y{:}E)dE \tag{46}$$

Thus, if h_s is known, we can compute P by deconvolving U with the inverse filter in a similar fashion to the veiling glare case. It should be noted that the overall video image $V(x,y)$ can also be rewritten as

$$V(x,y) = h_{VG}{*}{*}(h_s{*}{*}P)_{(x,y)} \tag{47}$$

where h_{VG} is the veiling glare PSF discussed in the previous section. Since the convolution operations commute, the equation can also be expressed as

$$V(x,y) = [P{*}{*}(h_{VG}{*}{*}h_s)]_{(x,y)} \tag{48}$$

The expression given in Eq. (48) indicates that if h_{VG} and h_s are known, they can be convolved to yield a two-dimensional overall PSF which can then be used to compute $P(x,y)$, the image which is degraded by beam hardening alone.

An earlier work performed by Shaw *et al.*[11] utilized an overall inverse filter which is given by

$$f(x,y) = \delta(x)\delta(y) - C\,{}^2\Pi(ax,ay) \tag{49}$$

where $^2\Pi(ax,ay)$ is a two-dimensional rectangle function[19] which is defined by

$$^2\Pi(ax,ay) = \begin{cases} 1 & |x|,|y| \leq 1/2a \\ 0 & \text{elsewhere} \end{cases} \qquad (50)$$

In Eq. (49), C is a constant to be adjusted. The application of the filter defined in Eq. (49) results in the high pass filtering operation of the degraded video image $V(x,y)$. Even though the filter used by these investigators[11] was quite arbitrary, it has been shown that the deconvolution of f linearizes the nonlinear videodensitometric volume curve quite successfully.

To conclude this section, we will simply state that the deconvolution techniques appear to be promising for obtaining accurate videodensitometric data. Current research at the University of California, Irvine, is focused on the determination of h_s and experimental validation of the approximations made in the development of the formalism.

5.3. Beam Hardening

In this subsection, we will assume that two correction techniques discussed in Sections 5.1 and 5.2 have been successfully applied to the acquired images. Thus, the processed images would only be affected by the beam-hardening artifacts discussed earlier. An obvious method for reducing the effect of beam hardening is to prefilter the X-ray beam heavily so that the X-ray beam becomes quasi-monoenergetic. The most common filters used for this purpose are aluminum or copper. Even though filtration of the X-ray beam reduces the beam-hardening effects, it also results in an increase in the quantum noise within the images. Thus, excessive filtration may prove to be undesirable since we cannot increase the X-ray tube output indefinitely. The approach which will be discussed in this section is a combination of tube filtration and the use of a lookup table. Since both veiling glare and X-ray scatter have been reduced by the use of methods discussed in Sections 5.1 and 5.2, we will only be dealing with the polychromatic primary X-ray beam. The polychromatic video signal given in Eq. (17) can be rewritten more explicitly as

$$V(x,y) = \beta \int D(x,y{:}E)N_0(x,y{:}E)\exp[-\int \mu(x,y,z{:}E)dz]dE \qquad (51)$$

where we assumed a parallel X-ray beam propagating along the z direction. The X-ray projection Pr at a location (x,y) can be written as

$$\Pr(x,y{:}L_T,L_I) = \ln[V(x,y{:}L_T,L_I = 0)/V(x,y{:}L_T - L_I,L_I)] \qquad (52)$$

In Eq. (52), it was assumed that we have a tissue thickness of L_T and $L_T - L_I$ before and after iodine injection. The detailed form of Eq. (52) is given by

$$\Pr(L_T, L_I) = \ln [I_0/I] \tag{53}$$

with

$$I_0 = \int dE\, D(E)\, N_0(E) \exp[-\mu_T(E) L_T] \tag{54a}$$

and

$$I = \int dE\, D(E)\, N_0(E) \exp\{-\mu_T(E) L_T - [\mu_I(E) - \mu_T(E)] L_I\} \tag{54b}$$

In Eqs. (53) and (54), we suppressed the dependence on (x,y) and μ_T, μ_I are the linear attenuation coefficients for tissue and iodine, respectively. In the monoenergetic limit, i.e., $N_0(E) = N_0\delta(E - E_0)$, Eq. (53) reduces to the more familiar form

$$\Pr_{mono}(L_T, L_I) = [\mu_I(E_0) - \mu_T(E_0)] L_I \tag{55}$$

In Figure 7, the polychromatic projection, Eq. (53), was plotted for two different tube filtrations. It should be noticed that in Eqs. (53) and (54), the constant amount of tissue along the beam acts as additional tube filtration, thus reducing the beam-hardening errors. This is similar to the constant-path water bath method which was used in the first-generation CT scanners.[20] We also notice that in the case of mask-subtracted videodensitometry, the projection given in Eq. (53) does not depend strongly on the tissue thickness L_T. In the monochromatic limit, there is no dependence on L_T. In order to generate a lookup table, we measured the primary projection using 4-mm aluminum filtration, 20 mA at 70 kVp. During various measurements, we used a different amount of "tissue" material and varied the concentration of 2-cm-thick Renografin-76. Some of the constant "tissue" materials experimentally used were: 0-cm lucite, 15-cm lucite, and 10-cm lucite plus 1-cm aluminum. The average polychromatic projection as a function of percent iodine concentration is given in Figure 28. It is observed that use of different types of tissue background material does not cause a significant spread in the projection data, i.e., the error bars for various concentrations of Renografin-76. The dashed straight line in Figure 28 was obtained by connecting the first three points. It represents the monochromatic projection given in Eq. (55). In order to

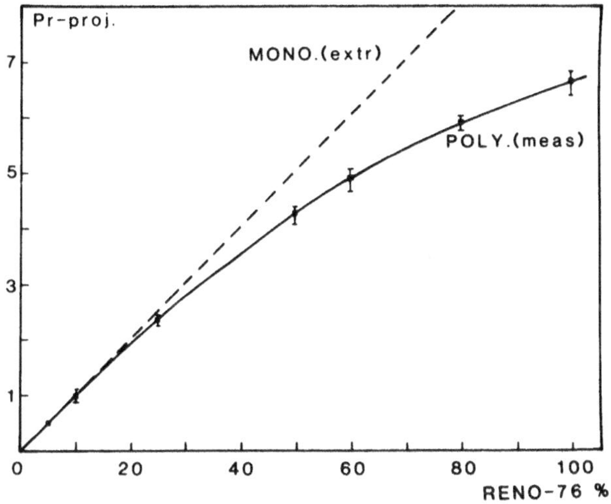

Figure 28. Measured polychromatic average projection used in the lookup function and extrapolated monochromatic projection as a function of iodine concentration.

make the experimentally measured projection data monochromatic, a transformation function was defined by

$$\text{Pr(mono)} = \sum_{j=1}^{M} a_j[\text{Pr(poly)}]^j \tag{56}$$

The coefficients a_j were determined by a least-squares fitting of Eq. (56) to the data shown in Figure 28. When $M = 5$ was used, i.e., a fifth-order polynomial, the coefficients shown in Table 2 were obtained. The lookup function obtained by using the five coefficients given in Table 2 is shown in Figure 29. Any polychromatic projection measurement along the ab-

TABLE 2. Coefficients of Lookup Function (Eq. 56)

j	a_j
1	1.375
2	−0.828
3	0.511
4	−0.109
5	0.774×10^{-2}

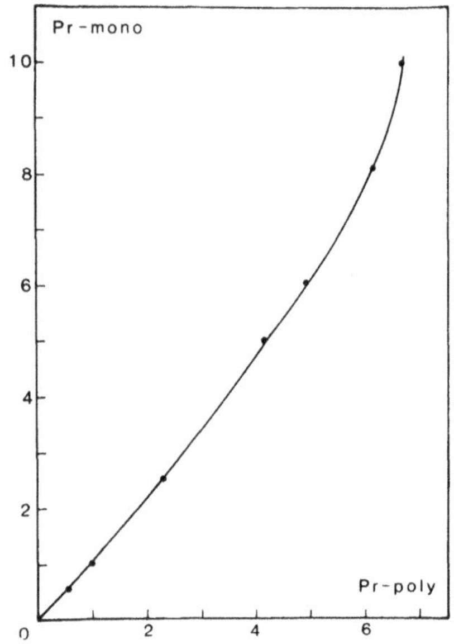

Figure 29. Monochromatic projection versus polychromatic one for 4-mm aluminum filtration, i.e., lookup function.

scissa may be converted to a monochromatic one by using Eq. (56) or Figure 29. The deviation from linearity in Figure 29 occurs more often for large values of Pr(poly). We used the experimental primary projection data which were discussed in an earlier section, and mathematically simulated the ellipsoid images as before. Each image was corrected, pixel by pixel, using the lookup function defined in Eq. (56). The measured volumes before and after the correction for LAO-view ellipsoids filled with 10% Renografin-76 are shown in Figure 30. The application of the lookup function (Eq. 56) with the coefficients given in Table 2 seems to improve the accuracy of volume measurements substantially. We should emphasize that the correction method discussed here is only applicable to volume measurements using a contrast agent since it is based on the presence of a constant thickness of background tissue and the displacement of blood by a contrast agent. The lookup function has to be precalculated using various thicknesses of tissue with the same amount of tube filtration that would be used in actual studies. Even though the correction appears to work well, the actual amount of correction would depend on the specific case under study.

6. APPLICATIONS

It was shown in the previous sections that measurements performed using II–TV-based digital X-ray imaging suffer from various degrees of

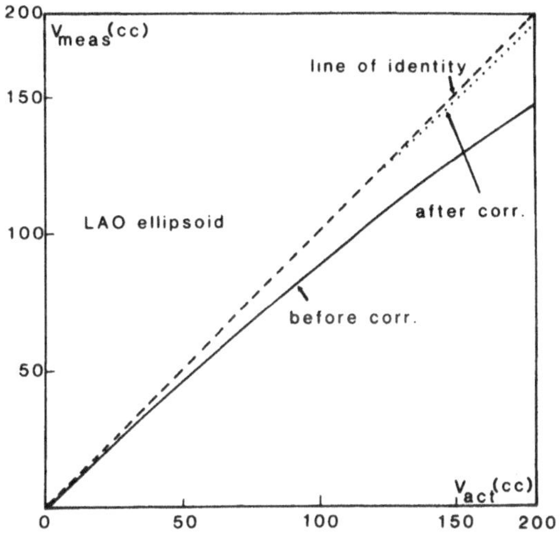

Figure 30. Measured versus actual volume of LAO ellipsoids before and after beam-hardening correction (simulation study).

errors due to the problems discussed earlier. In this section, applications regarding absolute and relative volume measurements without any correction techniques will be discussed.

6.1. Relative Volume Measurements

Here again, we will subdivide the subject into two areas. The first area will be the measurement of relative volumes of small objects. An important medical application is in the measurement of percent stenosis of vessels. The second application will be in the determination of larger relative volumes. In this case we will be concerned with the measurement of ejection fraction. The reason for dividing the applications into small and large volumes is shown in Figure 31. The actual volume of a balloon filled with 10% Renografin-76 is along the horizontal axis. The vertical axis is the videodensitometrically measured volumes in cubic centimeters. The normalization factor for actual volumes along the y axis was obtained by using the two lowest data points. The images were obtained at 70 kVp with 4-mm aluminum filtration and an X-ray grid. An interesting observation which can be made from Figure 31 is that even though the experimental curve deviates from a straight line, in two asymptotic regions there appears to be some linear response. These are marked with dashed lines in the figure. A similar observation was made by Shaw *et al.*[11] We notice that for small volumes (Region I), i.e., less than 40 cm^3, the curve appears to be a straight line with a zero intercept. However, in the second

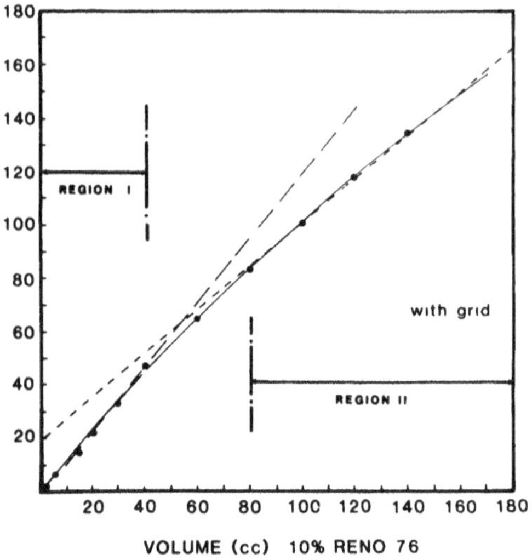

Figure 31. Measured versus actual volume of balloons filled with 10% Renografin-76 solution. Regions I and II are discussed in the text.

region, i.e., volumes larger than 80 cm^3, despite the fact that the curve seems to be a straight line, it has a nonzero intercept indicating that the measured volumes are systematically underestimated. As a result of these two linear regions, we predict that measurements done with small volumes would not suffer substantially from the errors mentioned earlier. The projection data presented in earlier sections also support this prediction. Thus, if one is interested in measuring small volumes such as vessels, one does not have to worry about applying the correction methods discussed in Section 5. However, if we were to perform measurements in Region II, the measured volumes would be underestimated by varying amounts. In order to determine the correct volumes, we would have to apply corrections to the acquired data. Even though we have identified two regions of linearity, it should be understood that the actual limits of these regions may vary depending on the experimental conditions. As a rule of thumb, for small volumes such as stenosis determination, one does not have to worry about corrections. On the other hand, the computation of ventricular ejection fraction (i.e., large volumes) would necessitate the correction of data prior to measurements.

6.1.1. Measurement of Ventricular Ejection Fraction

Percent of ejection fraction (EF) is defined as

$$\% \text{ EF} = 100(1 - V_{\text{ES}}/V_{\text{ED}}) \tag{57}$$

where V_{ED} and V_{ES} are the end-diastolic and end-systolic volumes. The computation of EF was done using the uncorrected images of balloons filled with 10% Renografin-76. The data collection was performed as follows. A mask image of the anthropomorphic chest phantom[21] was taken and stored in the computer memory. Another image was taken of a balloon filled with 10% Renografin-76 placed within the chest cavity. Two images were logarithmically subtracted without any other corrections. The process was repeated for balloons of different volumes. A sample subtracted image is shown in Figure 32. An ROI covering the object was drawn and the gray levels within the ROI were summed. The integrated value of gray levels is supposed to be a number proportional to the value of iodine mixture within the balloon. The points in Figure 31 were obtained by using these data. In order to compute various values of EF, several end-diastolic values were chosen and volumes smaller than that taken to be end-systolic volumes. The results for three different values of ED volumes are shown in Figure 33. As predicted earlier, most of the measured values are less than the actual ones. For comparison, we also show the line of identity in Figure 33. The amount of errors in this case appears to be less than that for the results presented in Section 4.2. The reason for the discrepancy is due to subtraction of a second background from logarithmically subtracted images. The secondary background in this case was chosen to be the difference between ED and ES ROIs, e.g., see Figure 34 for $V_{ED} = 160\ cm^3$ and $V_{ES} = 80\ cm^3$. The second background subtraction results in the cancellation of some of the errors in an arbitrary way. Similar results were reported by us earlier.[6]

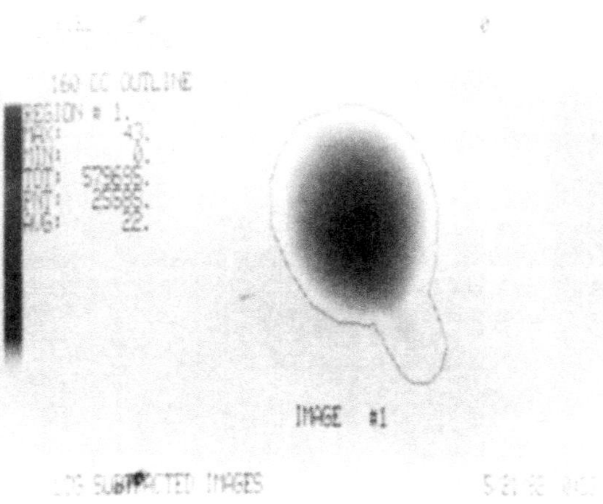

Figure 32. Selection of an ROI around $160\text{-}cm^3$ balloon for volume measurement.

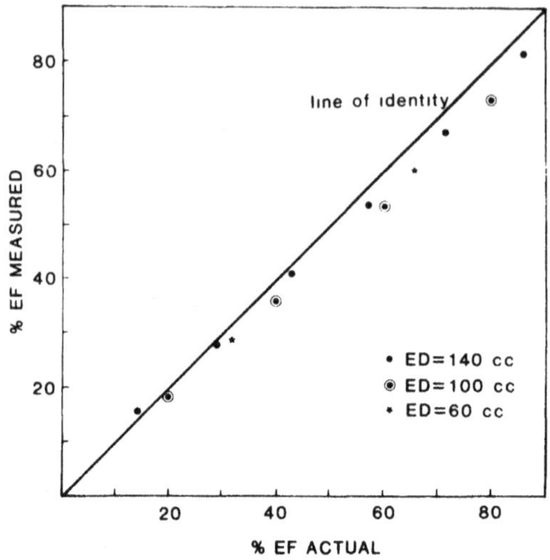

Figure 33. Measured versus actual percent EF using the balloon data.

6.1.2. Stenosis Measurement

The percent area stenosis is defined as

$$S_A = 100(1 - A_s/A_p) \tag{58}$$

where A_s and A_p are the cross-sectional areas of the stenotic and patent vessels, respectively. The data collection was performed by acquiring a

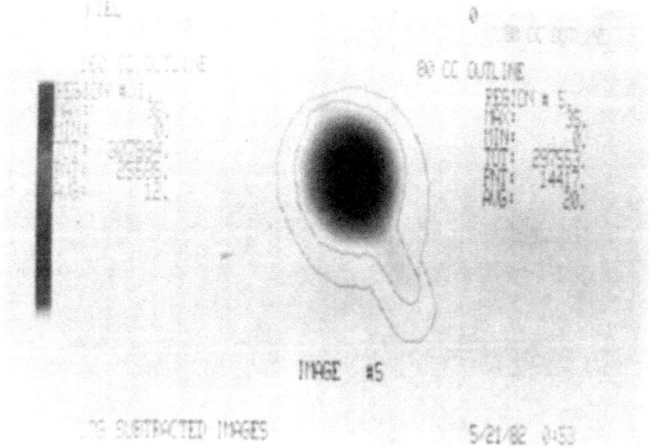

Figure 34. Selection of an ROI for secondary background subtraction.

mask image of the chest phantom. A second image containing the chest phantom and a Plexiglas block with drilled holes in it was taken. The holes had diameters ranging from 2 to 4.5 mm with 0.5-mm increments and were filled with 100% Renografin-76. The X-ray tube was operated at 75 kVp. After logarithmically subtracting the mask image from the one containing iodine, a rectangular ROI perpendicular to the midline of the patent section was chosen. The integrated digital numbers within the patent region I_p is given by

$$I_p = kV_p = kH_pA_p \tag{59}$$

where k is a constant of proportionality, H_p and A_p are the height of the ROI and cross section of the patent segment, respectively. In writing Eq. (59), we used the fact that volume $V_p = H_pA_p$. Similarly for the stenotic region, the integrated stenotic signal I_s is given by

$$I_s = kV_s = kH_sA_s \tag{60}$$

The computation of S_A is done by substituting Eqs. (59) and (60) in Eq. (58), thus resulting in

$$S_A = 100[1 - (I_sH_p/I_pH_s)] \tag{61}$$

The computation procedure is shown in Figure 35. The measured versus actual percent area stenosis is shown in Figure 36. It is seen that the largest errors occur for small values of S_A, i.e., normals where it is not

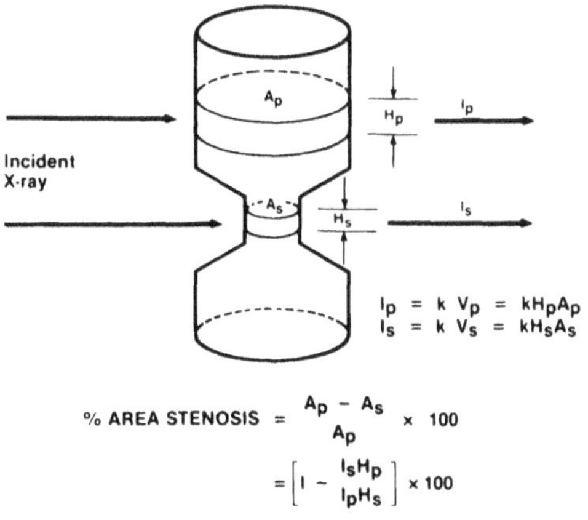

Figure 35. Stenosis measurement.

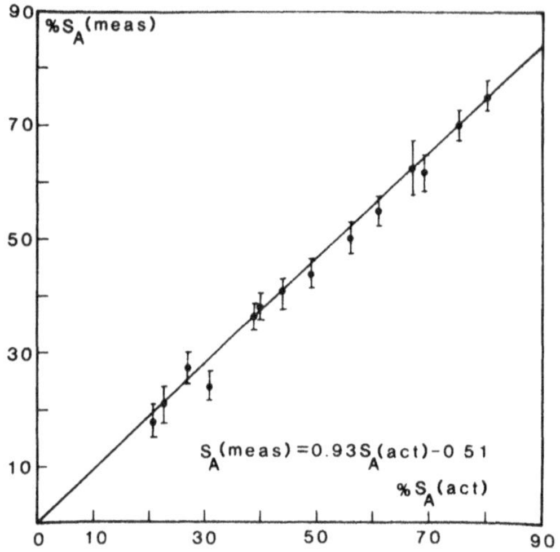

Figure 36. Measured versus actual percent area stenosis.

clinically significant. In any event, the measured values correlate quite well with the actual ones with a very small offset value. We thus conclude that the measurement of percent stenosis, i.e., relative volume of small objects, can be done videodensitometrically quite accurately without any corrections.

6.2. Absolute Volume Measurements

As we have seen before, the videodensitometric volume measurements are performed by integrating the gray levels within an ROI which encompasses the two-dimensional projection of the volume under consideration. Mathematically, the integration of gray levels within the ROI is expressed by

$$I = \int_{(x,y)} \int_{\epsilon \text{ ROI}} [\ln V_{\text{m}}(x,y) - \ln V_{\text{I}}(x,y)] dxdy \tag{62}$$

where V_{m} and V_{I} are the digital X-ray images before and after the introduction of the contrast agent into the volume of interest. We also point out that the mask image V_{m} is logarithmically subtracted pixel by pixel from the iodinated one V_{I}. The integrated signal I given in Eq. (62) actually is only proportional to the volume if the errors discussed in earlier sections are not serious. We can express this relationship as

$$I = k \cdot \text{volume} \tag{63}$$

where k is a constant of proportionality. Thus, absolute volume measurement using videodensitometric data (I) requires the determination of k. Kruger proposed a method[22] for the determination of k for small vessels by combining videodensitometric data with geometric assumptions about the object. A similar method was proposed by Henry and co-workers[23] which again utilized geometrical with videodensitometric information for the determination of ventricular volume. According to this method, it is assured that for the left ventricle, a region approximately two-fifths from the base has circular symmetry around the long axis of the ventricle. This is a less stringent assumption than the one which is used in the area–length method of left ventricular volume determination[24] where the ventricle is assumed to be an ellipsoid of revolution. If such a circular region exists, it may be used to determine k. This is done as follows. Let us assume that the circular region has a diameter D. If we pick a cylindrical slab of height H, then the geometrical volume of the slab is given by

$$V_0 = 0.25\pi HD^2 \tag{64}$$

We also need the videodensitometric signal from the same slab; let us call this I_0. According to Eq. (63), we have

$$I_0 = kV_0 \tag{65}$$

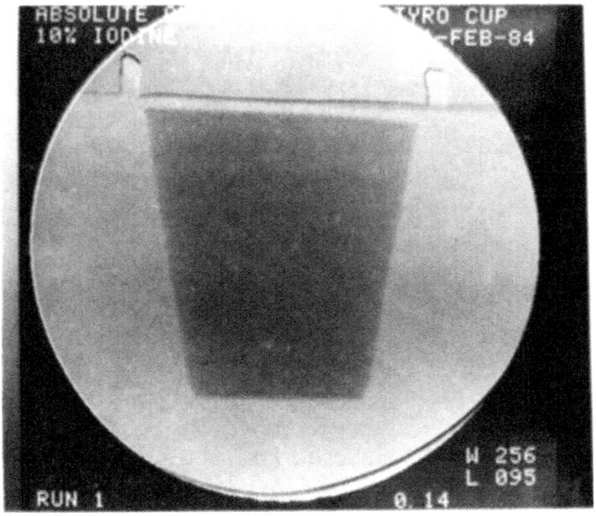

Figure 37. Subtracted Styrofoam cup filled with 10% Renografin-76 for absolute volume measurement.

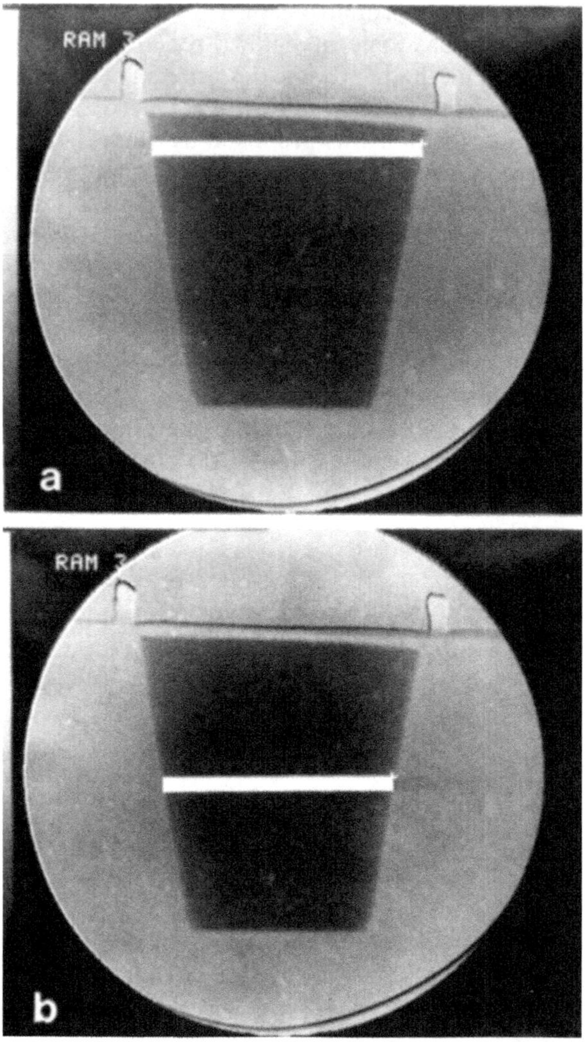

Figure 38. Selection of calibration ROIs: (a) top; (b) middle; (c) bottom.

which can be solved for k since we know both I_0 and V_0. By integrating the gray levels within the whole left ventricle, we obtain I_{LV} which is related to the ventricular volume by Eq. (63). Thus, we have

$$I_{LV} = kV_{LV} \tag{66}$$

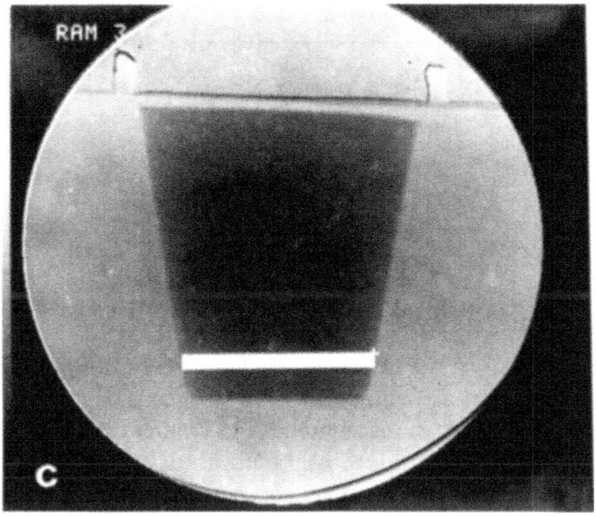

Figure 38. *(continued)*

Assuming that the contrast material is uniformly distributed within the ventricle, i.e., same constant in Eqs. (65) and (66), we can eliminate k and solve for V_{LV}. We obtain

$$V_{LV} = I_{LV}(V_0/I_0) \tag{67}$$

Equation (67) yields the absolute volume of the ventricle. An experimental simulation of this method was undertaken by filling a Styrofoam cup with water first (mask image) and then with a 10% solution of Renografin-76.[25] The exposure was made at 75 kVp with 4-mm aluminum filtration. The mask image was logarithmically subtracted from the iodinated one and the difference image is shown in Figure 37. Even though the cup is perfectly symmetric and all the dimensions are known, where to pick the calibration ROI still remains ambiguous. In order to test the influence of the location of calibration ROI on the final result, we used three different calibration ROIs. These were top, middle, and bottom of the cup as shown in Figure 38a–c. As a final check, we also mathematically simulated a cup of the same dimensions using the experimental projection data which were discussed in earlier sections. The actual volume of Renografin-76 solution within the cup was 150 cm^3. The results are given in Table 3. Experimentally determined absolute volumes vary from 168 to 140 cm^3 depending on the location of the calibration ROI, representing a change of error from

TABLE 3. Videodensitometric Absolute Volume Determination

Location of calibration ROI	Experimental		Computer simulation	
	Volume (cm^3)	% error	Volume (cm^3)	% error
Top	167.8	+11.9	166.7	+11.1
Middle	151.0	+0.7	149.0	−0.7
Bottom	140.0	−6.7	131.4	−12.4

+11.9 to −6.7%. For this object, the calibration ROI location in the middle yields the most accurate result. The results obtained by simulation using the experimental projection data reflect a similar behavior. The conclusion is that for an object which has a variable thickness, the location of the calibration ROI is important. Of course, if the data had not suffered from the degradation factors discussed earlier, the location of the calibration ROI would not have been important. This was tested by using the corrected projection data in the simulation. The calculated volumes were 150.9 and 151.1 cm^3 corresponding to the three calibration ROIs, indicating the insensitivity of the location of calibration ROI as expected. The conclusion drawn from this experiment is that even though one may obtain accurate absolute volume measurements for large objects, accuracy strongly depends on the location of the calibration ROI. In a real case with a patient, one never knows how much error is made if the images are not corrected prior to performing measurements. In the case of small objects, i.e., vessels, the errors would be much smaller even without any corrections, which is the same as before.

7. SUMMARY

In this chapter, we have discussed the various physical factors which cause the videodensitometric measurements performed with II–TV-based digital X-ray systems to be inaccurate. Experimental and simulation results demonstrating the contribution of each degradation factor were presented. We have shown that the degradation factors in increasing order of importance are beam hardening, veiling glare, and X-ray scatter within the patient. Various techniques for reducing the errors were introduced and discussed. In developing correction techniques, we avoided *empirical methods* such as background subtraction, lead disk placement on the object for scatter correction, and so on, since these methods work in limited situations and are not based on any sound physical principle. We believe that empirical methods result in unpredictable results when the experimental conditions are altered.

It was shown that the correction methods presented here are quite accurate in reducing veiling glare and beam-hardening artifacts. A new technique was also introduced for scatter correction. An important observation was that the measurements of small volumes was affected less by degradation factors as compared to larger volumes, e.g., small vessels versus the heart. As pointed out, this difference is due to two regions of linearity, one for small and the other for large volumes. In the small volume region, the offset value of the straight line response curve was quite small, resulting in accurate volume measurements. For the case of large volumes, in spite of the straight line response, a large offset appears, causing all the videodensitometric measurements to be systematically underestimated. Claims made by various authors regarding the good linear correlation between the measured and actual EF values seem to ignore the systematic underestimation and instead focus only on linearity. Furthermore, when one attempts to confirm the accuracy of a method using patient data and other "gold standard" techniques, one can never be sure what the true value is. We believe that the accuracy of any method should be confirmed using realistic phantoms where many uncertainties are eliminated.

ACKNOWLEDGMENT. This work was supported in part by PHS Grants 1R01 CA35243-02 and 1R01 HL31440-01A1.

REFERENCES

1. N. A. Baily, Video techniques for x-ray imaging and data extraction from roentgenographic and fluoroscopic presentations, *Med. Phys.* **7**, 472–491 (1980).
2. B. G. Trenholm, D. A. Winter, D. Myrain, and E. L. Lansdown, *Med. Biol. Eng.* **10**, 163 (1972).
3. N. R. Silverman, Clinical videodensitometry, *Am. J. Roentgenol.* **114**, 814 (1972).
4. C. A. Mistretta, R. A. Kruger, T. L. Houk, S. J. Riederer, C. G. Shaw, D. Ergun, W. Kubal, A. B. Crummy, W. Ziebel, G. Rowe, W. Zarnstorff, and D. Flemming, Computerized fluoroscopy techniques for non-invasive cardiovascular imaging, *Proc. SPIE* **152**, 65–71 (1978).
5. C. A. Mistretta, in *The Physics of Medical Imaging: Recording System Measurements and Techniques* (A. G. Haus, ed.), Institute of Physics, New York (1979).
6. O. Nalcioglu, J. A. Seibert, W. W. Roeck, W. L. Henry, J. M. Tobis, and W. D. Johnston, Comparison of digital subtraction videodensitometry and area length method in the determination of left ventricular ejection fraction, *Proc. SPIE* **314**, 294–298 (1981).
7. R. A. Kruger, Time dependent subtraction imaging using computerized fluoroscopy, Ph.D. thesis, University of Wisconsin (1978).

8. M. Abramowitz and I. Stegun, *Handbook of Mathematical Functions*, Dover, New York (1964).
9. R. A. Kruger, C. A. Mistretta, and S. J. Riederer, Physical and technical aspects of computerized fluoroscopy difference imaging, *IEEE Trans. Nucl. Sci.* **28,** 205–215 (1981).
10. W. Hendee, *Medical Radiation Physics*, Year Book, Chicago (1970).
11. C. G. Shaw, D. L. Ergun, P. D. Myerowitz, M. S. Van Lysel, C. A. Mistretta, W. Zarnstorff, and A. B. Crummy, A technique of scatter and glare correction for videodensitometric studies in digital subtraction angiography, *Radiology* **142,** 209 (1982).
12. J. A. Seibert, O. Nalcioglu, and W. W. Roeck, A deconvolution technique for the improvement of contrast of image intensifiers, *Proc. SPIE* **314,** 310–318 (1981).
13. O. Nalcioglu, J. A. Seibert, J. M. Boone, Y. Wang, W. W. Roeck, W. L. Henry, J. M. Tobis, and W. D. Johnston, in *Digital Imaging in Cardiovascular Radiology* (P. H. Heintzen and R. Brennecke, eds.), Thieme, Stuttgart (1983).
14. J. A. Seibert, O. Nalcioglu, and W. W. Roeck, Removal of image intensifier veiling glare by mathematical deconvolution techniques, *Med. Phys.* **12,** 281 (1985).
15. J. A. Seibert, O. Nalcioglu, and W. W. Roeck, Characterization of the veiling glare PSF in x-ray image intensified fluoroscopy, *Med. Phys.* **11,** 172 (1984).
16. B. R. Frieden, in *Picture Processing and Digital Filtering* (T. S. Huang, ed.), Springer-Verlag, Berlin (1975).
17. E. O. Brigham, *The Fast Fourier Transform*, Prentice–Hall, Englewood Cliffs, N.J. (1974).
18. H. H. Barrett and W. Swindell, *Radiological Imaging*, Academic Press, New York (1981).
19. R. Bracewell, *The Fourier Transform and Its Applications*, McGraw–Hill, New York (1965).
20. E. C. McCullough, Photon attenuation in computer tomography, *Med. Phys.* **2,** 307 (1975).
21. J. G. Pearce, E. N. C. Milne, G. D. Gillan, and W. Roeck, Development of a radiographic chest phantom with disease simulation, *Invest. Radiol.* **14,** 181 (1979).
22. R. A. Kruger, Estimation of the diameter of and iodine concentration within blood vessels using digital radiography devices, *Med. Phys.* **8,** 652 (1981).
23. J. M. Tobis, Y. Wang, O. Nalcioglu, and W. Henry, Absolute left ventricular volume obtained by videodensitometry and DSA, *J. Am. Col. Cardiol.* **2,** 616 (1983).
24. H. T. Dodge, H. Sandler, D. Ballew, and J. D. Lord, The use of biplane angiocardiography for measurement of left ventricular volume in man, *Am. Heart J.* **60,** 762 (1960).
25. J. A. Seibert, O. Nalcioglu, J. Tobis, W. Roeck, and W. Henry, Factors influencing quantitation of absolute ventricular volume by digital subtraction angiography, in press.

6

Recursive Filtering Techniques Applied to Digital Subtraction Angiography

R. A. Kruger

1. INTRODUCTION

The most successful application of digital radiography to date has been digital subtraction angiography (DSA). Its main clinical use has been in the evaluation of head, neck, and abdominal arteries. Mask-mode image subtraction has been the most widespread DSA technique.[1–5] For non-cardiac studies, this technique uses X-ray exposures made approximately once per second for a total of 15 to 20 exposures spanning the duration of the arrival and washout of the contrast material bolus. Typically, exposure times are in the range of 20 to 200 msec, depending on anatomy. Of the resultant images, two are subtracted from each other to form the final difference image, one taken from among those containing little or no contrast, and another taken at or near peak opacification. Because the two images are acquired at different times, the process is called temporal subtraction.

Integrated remasking,[6] matched filtering,[7–10] and recursive filtering[11–15] all are methods in which some process equivalent to temporal subtraction is used to form an image resembling a mask-mode result. The major difference between mask-mode and these three methods lies in the number of images that are used to form each difference image. With mask mode the effective integration time for each image (during which image blurring may occur) is that of the X-ray pulse width, 20 to 200 msec. X-

R. A. Kruger • Department of Radiology, University of Utah College of Medicine, Salt Lake City, Utah 84132.

ray dose utilization, however, is very poor, since most of the images (and their associated X-ray exposures) are discarded. In contrast to mask subtraction, these three modifications all use X-ray exposures that extend over 1 to 10 sec or more. The potential for image blurring over such a period may be offset by a greater utilization of patient exposure and a corresponding lowering of image noise, or a lowering of patient exposure without increasing image noise compared with mask-mode subtraction.

This chapter deals with some of the properties and applications of recursive image sequence processing for use in DSA.

2. TEMPORAL FILTERING THEORY

First, it should be realized that both mask subtraction and recursive filtration are specific expressions of a general linear filtering approach. In either case a sequence of images (I_i) is acquired and stored. The subscript i denotes the time at which each image was acquired. The rate at which images were acquired is immaterial.

Figure 1 shows an example where 25 images were acquired at a rate of 2 images/sec spanning the flow of contrast material through an artery of interest. This contrast dilution curve describes the temporal variation of vessel image contrast which typically results from an intravenous injection of contrast material. This dynamic change represents the "signal" which is to be isolated by separating or "filtering" it from other types of temporal image variations.

A general linear filter applied to this set of images forms a filtered image, I, where I is a linear combination of the image samples, i.e.,

$$I = \sum k_i I_i \tag{1}$$

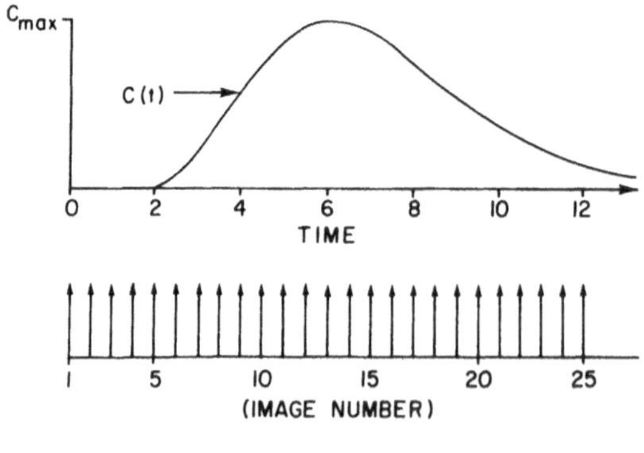

Figure 1. All DSA imaging procedures require acquisition of a set of images spanning the time of arterial opacification. During this time, arterial opacification varies in time as represented by the iodine concentration dilution curve, $C(t)$. In this example, 25 images are collected at a rate of 2 per second.

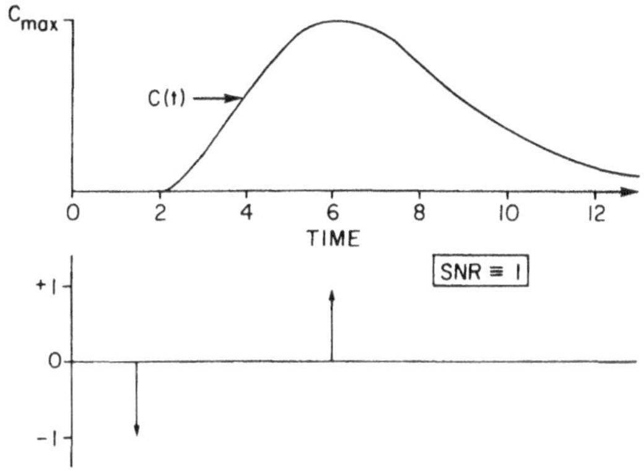

Figure 2. For mask-mode subtraction, 2 of the 25 images are retained and subtracted from one another, in this case $I_{13} - I_4$. This process is equivalent to applying a linear filter with the coefficients +1 and −1, as shown.

For angiographic purposes we are interested only in those linear filters for which the following condition is valid:

$$\sum_i k_i = 0 \tag{2}$$

This criterion ensures that stationary background anatomy is removed.

Any processing scheme that meets the criteria expressed in Eqs. (1) and (2) will provide some measure of success in isolating vascular anatomy following bolus injection of contrast material. The set of weighting coefficients (k_i) defines the properties of each filter.

As an example, a choice of mask-mode subtraction filter coefficients is shown in Figure 2. In this case two images are retained for processing, I_4 and I_{13}. These two images are assigned weighting coefficients of −1 and +1, respectively. The other images, which are not used, are assigned a zero coefficient. Image I_4 is called the "mask" and image I_{13} is the post-vessel opacification image chosen near peak opacification. In general, mask subtraction can be thought of as that filter defined by the following set of coefficients:

$$K_l = -1; \qquad K_m = 1; \qquad K_i(\neq l,m) = 0 \tag{3}$$

Image l serves as the mask and image m is chosen near peak opacification. The other images are discarded.

For the recursive filters considered in this chapter, all the weighting coefficients are nonzero, i.e., all of the images are retained for processing. For noncardiac imaging the weighting coefficients shown in Figure 3 have proven to be of clinical value. Some of the coefficients are positive and

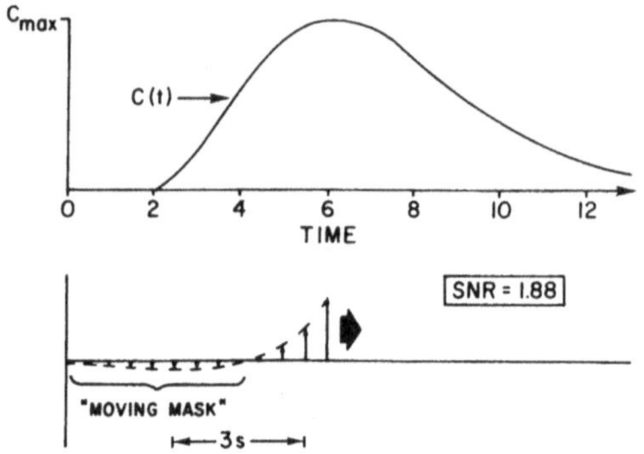

Figure 3. Recursive filter coefficients which have proven to be useful for intravenous and intraarterial angiography.

some are negative with the sum of all coefficients equal to zero as required by Eq. (2). The mean time between images assigned positive coefficients and those assigned negative coefficients is approximately 3 sec. Therefore, image changes which occur over a time scale on the order of 3 sec are "isolated" by such a filter.

The recursive filter only operates on images acquired up to a certain time. A new filtered image is formed after each additional image acquisition. The weighting coefficients simply are shifted to the right (arrow, Figure 3) after each image acquisition. Consequently, an image sequence results. Furthermore, such a filtering technique may be applied during image acquisition, i.e., in real time.

Because a recursive filtering technique must fulfill the criterion expressed in Eq. (2), it is possible to view this technique as a subtraction process. Because $\Sigma k_i = 0$, some values of k_i are positive and some are

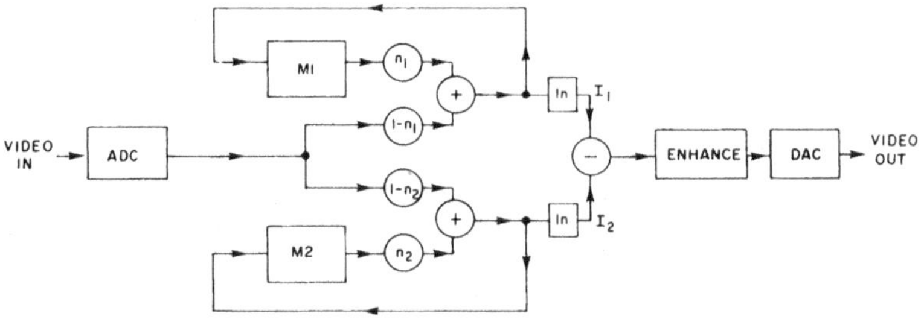

Figure 4. Approximate band pass filter. Real-time recursive video image processor used at the University of Utah. Video images are digitized in real time (30 frames/sec) during continuous X-ray exposure. The processor contains two digital memory planes (each 480 × 792 × 16 bits), scaling circuitry (n_1,n_2), logarithmic lookup tables (ln), and subtraction and enhancement circuitry.

negative. This being the case, we can rewrite Eq. (1) as

$$I = \sum_i |k_i(>0)|\, I_i - \sum_i |k_i(<0)|\, I_i \tag{4}$$

The negative coefficients have been written as $-|k_i|$ so that the subtraction process is expressed explicitly. Recursive filtering then can be thought of as a combination of image integration and subtraction.

The form of Eq. (4) also suggests the form of the processing architecture which can be used to implement a recursive filter for DSA. The image processor architecture which has been developed at the University of Utah is shown in Figure 4. There exist two image memories for storing positive (k_1) and negative (k_2) coefficients. The parameters n_1 and n_2 determine the extent of image integration as well as the effective time separation between positive and negative images. The image acquisition and display rates have been chosen to be 30 per second, consistent with conventional video scanning procedures.

Since many more than two images are used with recursive processing, better use of the X-ray exposure can be realized compared to mask-mode subtraction where only two images are used. Typically, a factor of ap-

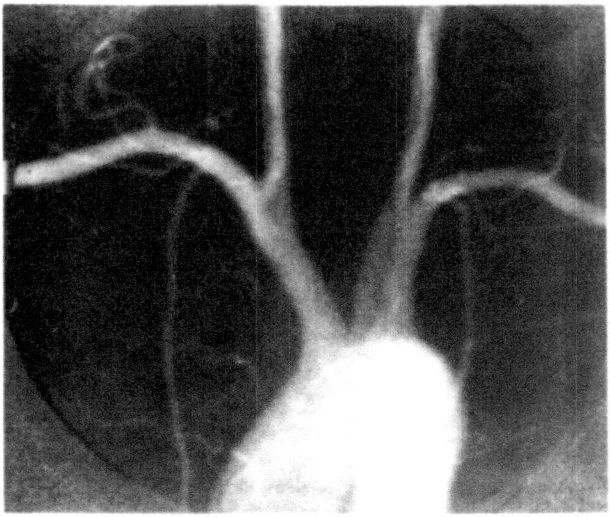

Figure 5. Intravenous DSA image formed using the recursive processor shown in Figure 4 and filter coefficients shown in Figure 3. Forty milliliters of 76% contrast material was injected into the superior vena cava at a rate of 20 ml/sec (60 kVp, 5 mA). A good view of the aortic arch in an RPO projection is seen.

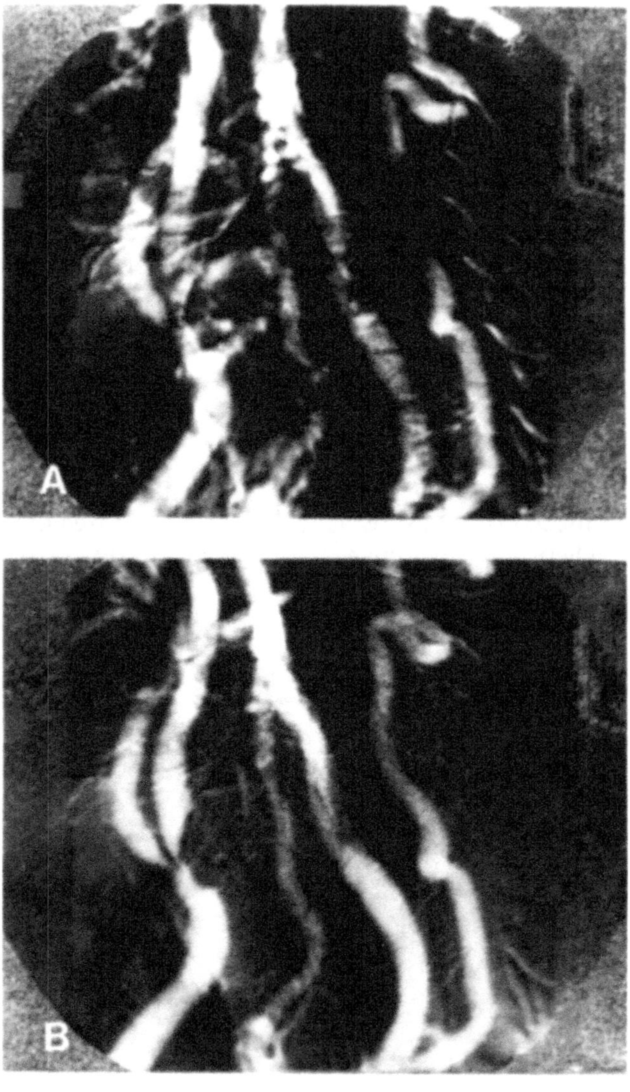

Figure 6. (A) Image of the carotid arteries using an identical procedure to that used to obtain Figure 5 except that the tube current was increased to 10 mA. Significant movement artifact is evident. (B) Filtering the image sequence a second time, after editing, removes most of the motion artifact.

proximately 2 less image noise results when using recursive filtering as opposed to mask-mode subtraction for the same patient X-ray exposure. For the example shown in Figure 3, a factor of 1.88 noise reduction is predicted. Alternatively, less patient exposure (approximately fourfold reduction) can be achieved without increased image noise compared to mask-mode subtraction. Reductions in the quantity of injected contrast dye also have been reported[10] using a technique analogous to recursive filtering.

3. NONCARDIAC CLINICAL RESULTS USING RECURSIVE FILTERING

Recursive video image processing has been used for two years at the University of Utah Medical Center with results comparable to those reported by other research groups using alternative approaches. Several notable differences in method should be pointed out:

1. Continuous rather than pulsed X-ray exposure is used.
2. Standard, 30 frame/sec video acquisition is used.
3. A standard R/F installation and average-quality TV camera are employed.

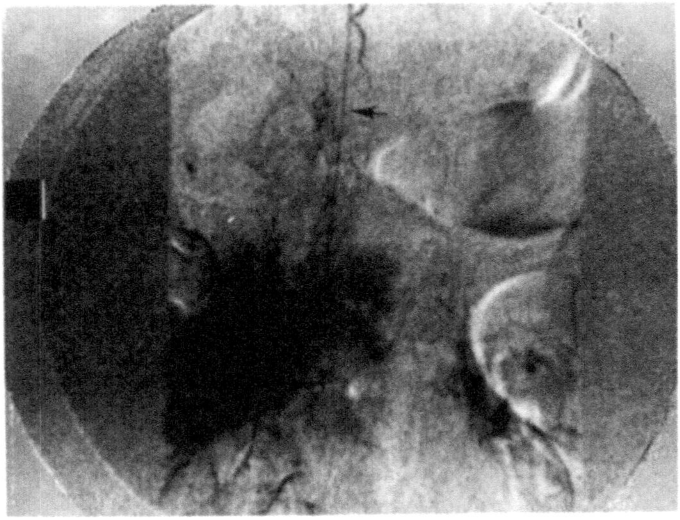

Figure 7. Intraarterial DSA using recursive filtering. Injection of 30% contrast material in an intercostal artery reveals the difficult-to-visualize artery of Adamkiewicz (arrow).

4. Tube currents of 1–30 mA are used typically.
5. Patient X-ray exposure is reduced at least fourfold compared to other reported techniques.

Both intravenous and intraarterial studies have been performed routinely. Examples of these studies are shown in Figures 5–7. The examples chosen indicate several features of this type of extensive image averaging technique:

1. Vessel pulsation due to cardiac activity does *not* result in significantly blurred arterial walls in most arteries. Exceptions are the arteries proximal to the heart. The aortic arch, however, still images sharply (Figure 5).
2. Patient motion such as swallowing is handled in a manner analogous to remasking techniques. By editing and refiltering a once filtered image sequence, patient motion artifact can be reduced significantly (compare Figure 6A and B).
3. Fine vessel resolution and low contrast detection are comparable to other techniques (Figure 7).

4. CARDIAC APPLICATIONS

Rather than describing a filter in terms of its weighting coefficients, it often makes more sense to characterize the response of a filter to

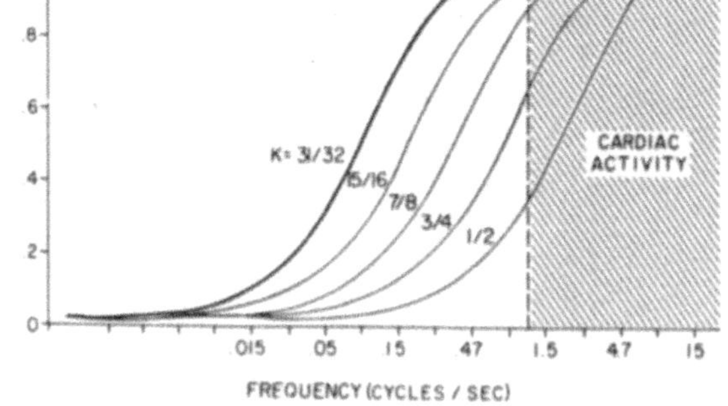

Figure 8. Temporal frequency responses of a number of candidate recursive filter functions which might be considered for cardiac imaging.

sinusoidal temporal changes which oscillate at various frequencies. Any time-varying signal (image sequence) can be described by a combination of such pure sinusoidal variations. A filter's frequency response is defined as follows. Assume a pure sinusoidal frequency is operated on by a particular filter. The effect of the filter would be to alter the amplitude (intensity) of the sinusoid but not its frequency. The ratio of the filtered to the unfiltered sinusoidal amplitude is the filter response at that frequency. This ratio will be different at different frequencies. The filter's responses at all possible frequencies describe the complete frequency response of the filter.

The concept of frequency response is perhaps easiest understood in the context of imaging rapidly moving cardiac structures such as the left ventricle or coronary arteries. The motion of the heart is approximately periodic with a fundamental frequency identical to the heart rate. For a heart rate of 60 beats/min the fundamental cardiac frequency is 1 cycle/sec. The image variation due to these moving structures is best characterized by temporal frequencies above 1 cycle/sec. These frequencies are shown graphically in Figure 8. Also shown are several frequency responses which can be formed with the dual memory recursive filter shown in Figure 4, and which might be considered for cardiac imaging.

The reason these filter responses are good candidates for cardiac

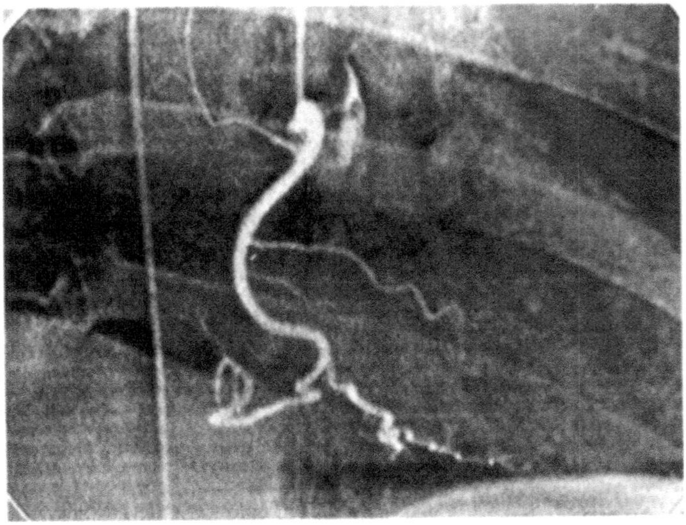

Figure 9. Single frame of a human right coronary artery using the filter response labeled as 15/16 in Figure 8.

imaging is that their responses in the region of cardiac activity (> 1 cycle/sec) are high, and at the same time the responses below this frequency are low (zero for stationary structures). Because these filter responses can be realized with recursive processing, a filtered image sequence primarily showing cardiac motion can be formed, in this case at a rate of 30 images/sec.

The parameters (n_1, n_2) determine the filter's exact response. The combination (31/32,0) has been found useful for left ventriculography. For selective coronary angiography the combination (15/16,0) permits "panning" of the image intensifier and yet still isolates the rapidly moving coronary arteries. Figure 9 shows a human right coronary artery imaged with such a recursive filter. Background suppression, though not perfect, is good.

Recursive filtering techniques have been applied successfully to virtually every artery in the body. These techniques can be viewed as special cases of subtraction imaging. These techniques complement, rather than compete with, other DSA techniques. Recursive filtering should therefore be viewed as another strategy for isolating and displaying radiographic image information.

REFERENCES

1. C. A. Mistretta, A. B. Crummy, and C. M. Strother, Digital angiography: A perspective, *Radiology* **139,** 273–276 (1981).
2. R. A. Kruger, C. M. Mistretta, T. L. Houk, S. J. Riederer, C. G. Shaw, M. M. Goodsitt, A. B. Crummy, W. Zwiebel, J. C. Lancaster, G. G. Rowe, and D. Fleming, Computerized fluoroscopy in real time for nonivasive visualization of the cardiovascular system, *Radiology* **130,** 49–57 (1979).
3. C. A. Strother, J. F. Sackett, A. B. Crummy, F. G. Lilleas, W. J. Zwiebel, W. D. Turnipseed, M. Javid, C. A. Mistretta, R. A. Kruger, D. L. Ergun, and C. G. Shaw, Clinical applications of computerized fluoroscopy: The extracranial carotid arteries, *Radiology* **136,** 781–783 (1980).
4. P. C. Christenson, T. W. Ovitt, H. D. Fisher, M. M. Frost, S. Nudelman, and H. Roehrig, Intravenous angiography using digital video subtraction: Intravenous cervicocerebrovascular angiography, *Am. J. Roentgenol.* **135,** 1145–1152 (1980).
5. T. F. Meaney, M. A. Weinstein, F. Buonocore, W. Pavlicek, G. P. Borkowski, J. H. Gallagher, B. Sufka, and W. J. MacIntyre, Digital subtraction angiography of the human cardiovascular system, *Am. J. Roentgenol.* **135,** 1153–1160 (1980).
6. D. R. Enzmann, W. T. Djang, S. J. Riederer, W. F. Collins, A. Hall, G. S. Keyes, and W. R. Brody, Low-dose high frame digital subtraction angiography, *Radiology* **146,** 669–676 (1983).
7. R. A. Kruger and P.-Y. Liu, Digital angiography using a matched filter, *IEEE Trans. Med. Imag.* **1,** 16–21 (1982).
8. S. J. Riederer, A. L. Hall, J. K. Maier, N. J. Pelc, and D. R. Enzmann, The technical

characteristics of matched filtering in digital subtraction angiography, *Med. Phys.* **10,** 209–217 (1983).

9. S. J. Riederer, D. R. Enzmann, A. L. Hall, N. J. Pelc, and W. T. Djang, The application of matched filtering to x-ray exposure reduction in digital subtraction angiography: Clinical results, *Radiology* **146,** 349–354 (1983).
10. S. J. Riederer, D. R. Enzmann, W. R. Brody, and A. L. Hall, The application of matched filtering to contrast dose reduction in digital subtraction angiography, *Radiology* **147,** 853–858 (1983).
11. R. A. Kruger, A method for time domain filtering using computerized fluoroscopy, *Med. Phys.* **8,** 466–469 (1981).
12. R. G. Gould, M. J. Lipton, P. Mengers, and R. Dahlberg, Investigations of a video frame averaging digital subtraction fluoroscopic system, *Proc. SPIE* **314,** 184–191 (1981).
13. R. A. Kruger, P.-Y. Liu, W. Bateman, and J. Nelson, Time domain filtering using computerized fluoroscopy: Intravenous angiography applications, *Proc. SPIE* **314,** 319–324 (1981).
14. J. A. Nelson, F. J. Miller, R. A. Kruger, P.-Y. Liu, and W. Bateman, Digital subtraction angiography using a temporal bandpass filter: Initial clinical results, *Radiology* **145,** 309–313 (1982).
15. R. A. Kruger, F. J. Miller, J. A. Nelson, P.-Y. Liu, and W. Bateman, Digital subtraction angiography using a temporal bandpass filter: Associated patient motion properties, *Radiology* **145,** 315–320 (1982).

7

Energy-Selective Radiography

A Review

L. A. Lehmann and R. E. Alvarez

1. INTRODUCTION

Energy-selective radiography makes use of body transmission measurements over multiple X-ray energy spectra. These measurements of the spectral properties of the transmitted radiation add a novel dimension to classical radiography, and provide more information than a measurement over a single arbitrary spectrum for the same radiographic exposure. The central questions are: How clinically relevant is this added information? How can measurements be combined to provide an improved image or to isolate valuable and previously unobtainable information? How can imaging systems be constructed to acquire energy-selective measurements? What are the noise properties of energy-selective images, and how do they compare to conventional images?

Energy-selective systems actually reduce the effects of X-ray quantum noise. These systems can image hard-to-detect features with higher signal-to-noise ratio (SNR) than can conventional systems using identical X-ray exposures.

Overlying anatomical detail also represents an important noise source in radiography. Energy-selective systems provide the viewer a display parameter that can arbitrarily control the relative contrast between any two materials of different effective atomic composition. The viewer may choose to remove at will any single material type from the image.

Energy-selective data make possible computed tomography (CT) reconstructions totally lacking in polyenergetic artifacts. In any properly constructed energy-selective system, CT or single projection, the resulting

L. A. Lehmann and R. E. Alvarez • DigiRad Corporation, Palo Alto, California 94303.

images are truly portable and system independent, being no longer functions of the source spectra employed. The framework of energy selectivity encourages quantification, analysis, and standardization. It is also possible, in an energy-selective system, to identify the approximate effective atomic number of an unknown medium; this type of material characterization has never before been possible with single-projection radiography.

Through the course of this review, a theoretical, and for the most part idealized approach will be taken to the study of information in energy-selective systems. The performance of energy-selective and conventional imaging systems will be compared, the requirements the system must meet will be indicated, and the unique features of energy-selective images will be analyzed.

We will begin with a short review of early contributions to energy-selective radiography, and then describe some of the physical forms of dual-energy apparatus. Section 3 presents a linear model of X-ray attenuation which is the theoretical basis for the analysis which follows. Section 4 questions the possibility of extracting line integrals of attenuation from ordinary transmission measurements. It establishes that conventional X-ray systems cannot measure these attenuation line-integrals without familiar polyenergetic nonlinearity artifacts. By measuring transmission over as few as two energy spectra, it is possible to accurately measure the line integral of the attenuation coefficient, eliminate the artifact, and extract "all" the energy-dependent behavior of the line integral of attenuation. Section 5 explores applications of energy-selective measurements, including the possibility of forming material-canceled images, of arbitrarily varying the contrast ratio of two substances of different effective atomic number, and of tissue characterization by X-rays. Contrast and noise are studied in detail in Section 6 and a comparison is made of contrast and noise in conventional and energy-selective imaging systems.

Early efforts in energy-selective radiography focused on improving the contrast of the radiographic contrast agent iodine. Iodine has an absorption edge or discontinuity in its attenuation coefficient at an energy of 33.155 keV. The attenuation coefficients of the elements composing body tissues are smooth functions of energy throughout the diagnostic region of 30–150 keV. Jacobson[1] constructed an imaging system which exploited the iodine absorption edge to enhance the contrast of iodinated structures. Two transmission measurements were made for every projection path, one at an energy just above the iodine K-absorption edge and one just below. The measurement difference is essentially zero where there is no iodine. This method provided a means of quantifying the iodine mass-thickness (g/cm^2) along a projection path. Jacobson used secondary target emission sources to generate the two monoenergetic beams.

In the years 1973–1977, Mistretta and others improved on the *K*-edge technique.(2–4) They recommended use of a conventional source, heavily filtered by iodine and cerium. Such a source is more efficient but less monoenergetic than secondary emission sources. Departure from true monoenergetic beams leads to an error in the suppression of bone and soft tissues, an error which can be corrected using a measurement at a third energy.(5–8) At about the same time, Macovski,(9,10) Alvarez,(11) and Yeh(6) proposed a method for spatially encoding the two measurements onto a single film, thereby resolving problems of film misregistration and making possible the extension of *K*-edge imaging to ordinary film detection.

Yet the absorption edge technique has not found its way into clinical practice, largely because there are few currently available sources of intense and nearly monoenergetic X-rays at the *K*-edge energy of iodine.(12) Of course, this method cannot enhance the contrast of natural body materials since their *K*-edge energies are well below the diagnostic range.

These difficulties, in connection with absorption edge imaging, led to exploration of multiple broadband measurements at energies above all absorption edges where the attenuation coefficients are smooth functions of energy. The early investigators were seeking some analogy to color imaging within the visible region. There was interest in the early 1950s in ad hoc assignment of images measured at different kVp's to the primary colors for exposure of color film. Such methods(13–15) distinguished biological materials of different atomic composition by their apparent color on the "color radiograph."

The present discourse considers only broadband, absorption edgeless imaging, a modality which makes most efficient use of X-ray sources and patient exposure. This presentation will summarize much of the analytical structure of broadband energy-selective radiography that dates from the mid-1970s.

2. APPARATUS FOR ENERGY-SELECTIVE IMAGING

Energy selectivity depends on the measurement of object transmission over multiple and different spectra. Some energy-selective radiographic systems view the object sequentially, while either varying the spectral content of the source or varying the spectral response of the detector. In other systems, the detector has inherent energy discrimination which, like a multichannel analyzer, measures both radiation flux and an estimate of the energy spectrum transmitted through the object during one exposure.

Stanford University, in collaboration with the General Electric Company,[16–18] constructed an energy-selective radiography system of the first type using a CT linear array detector of 512 elements and a source which was collimated to produce a planar fan-shaped beam. A typical scan lasted approximately 10 sec, while the patient moved through the stationary beam on a motorized table. For energy selectivity, the X-ray tube voltage alternated between 85 and 135 kVp at a cycle rate of 30 Hz, and the detector array measured the transmitted flux at a 60-Hz rate. In synchrony with tube voltage changes, the system switched source filters to further the spectral difference of the two measurements. A bronze filter hardened the high-energy beam somewhat, and an erbium filter softened (reduced the average energy) of the lower-energy beam by suppressing the tungsten target's characteristic radiation and other high-energy components of that beam while allowing spectral components just under the erbium *K*-absorption edge to pass. The introduction of erbium to the low-energy beam reduced the patient exposure to about one-third of its former value while actually increasing the SNR. Erbium improved the conditioning of the signal-processing operations and spared the patient the effects of moderate energy photons which add little information.

Line-scanned systems, such as the one at Stanford, are capable of high-dynamic-range measurements without errors due to scatter. The data acquisition system is extraordinarily linear and repeatable. Their detectors are X-ray quantum limited with high quantum detection efficiencies (QDE). As such, they are excellent tools for research and for quantitative studies. But line-scanned systems suffer from tube/collimation considerations which limit their spatial resolution and useful output flux. They are expensive, slow, and can introduce geometric distortions if the tissues move during the data collection period. Line-scanned systems are also inappropriate for fluoroscopic and other real-time applications.

Kruger[19] described a film-based energy-selective system which used voltage switching and sequential exposures. In the time between exposures, X-ray films were changed as might be done with an automated film changer. Kruger chose appropriate film gammas and used a photographic method to subtract the two initial films. This nonelectronic method for generation of material-canceled images provided surprisingly good results. The technique is not limited in the tube load, spatial resolution, or system cost, but the processing step is awkward. It is difficult to register the two radiographs accurately, the dynamic range is limited, and it is susceptible to artifacts if there is patient motion in the time between the two exposures.

There has also been a long history of systems using detectors with inherent energy sensitivity. Ritchings and Pullan[20] described a CT linear detector array in which alternate detectors were covered with a filtering window. The modification was simple, but the system was not dose efficient owing to the post-patient filtering. The multiple measurements were not valid for identical projection paths, thus leading to the possibility of inconsistent measurements at sharp object boundaries.

Alvarez and Macovski (U.S. Patent 4,029,936, 1977) described a dual-channel scintillation detector which is dose efficient, and which forms simultaneous high- and low-energy measurements for identical projection paths. Brooks and DiChiro[21] implemented this type of detector in a modified EMI Mark I CT scanner by overlapping two scintillation detectors, a thin CaF_2 in front, and a thick NaI scintillator behind. Energy-discriminating detectors such as these have the advantage that no filtration of the source is necessary, and in addition, the source can be operated at a high kVp. Both of these factors add efficiency to the source utilization which can be a limiting factor in linear array systems.

Fenster, of the University of Toronto,[22] built a two-channel energy-selective xenon gas detector. This detector was also fan-shaped and was used in linear array energy-selective CT.

Stone[23] demonstrated a novel pulse height analysis detector array constructed of high-purity germanium. The array was fast enough to operate at diagnostic photon rates and had an energy resolution effective rms noise of 25 keV.

Energy-selective detectors have also been built using film, which as mentioned previously enjoys high spatial resolution, temporal resolution, and excellent source utilization. In addition, film-based energy-selective systems can be fully compatible with existing X-ray equipment. The dual-energy cassette[11] contains multiple emulsions, specially chosen X-ray screens which act as sensitizers, and X-ray filters (Figure 1). Note that

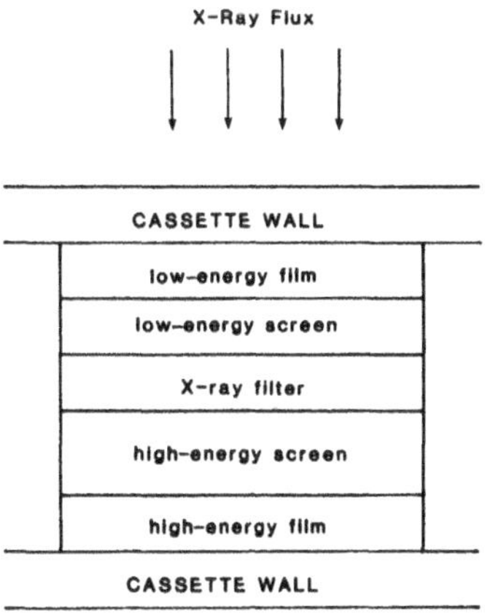

Figure 1. Construction of a dual-energy dual-film cassette.

two ordinary X-ray films are actually used in the cassette. DigiRad Corporation (Palo Alto, California) has commercialized a system and cassette constructed in this way.

This film method requires accurate registration of the two films. This is not as difficult a problem as it may seem at first glance, since both films are exposed simultaneously, and there is therefore no possibility of patient motion artifacts. The DigiRad cassette employs pins in the cassette and in the film scanner for alignment. Registration is, however, theoretically limited by the distance between the two screens and the consequential difference in magnification on the two films. Small misregistrations, even much smaller than the system resolution length, can have significant detrimental effects on energy-selective images of sharp or high-dynamic-range edges. The difference between high- and low-energy images is very subtle and the processing steps amplify small differences between the two.

A technique for eliminating such misregistration and for saving the expense and burden of a second film was outlined by Strul.[24] As shown in Figure 2, the single-film energy-selective cassette codes the energy information as the amplitude modulation of a grating image. The grating frequency is higher than the screen's passband but still within the film's modulation transfer function (Figure 2b). The arrangement uses space within the modulation transfer function of the film that was previously

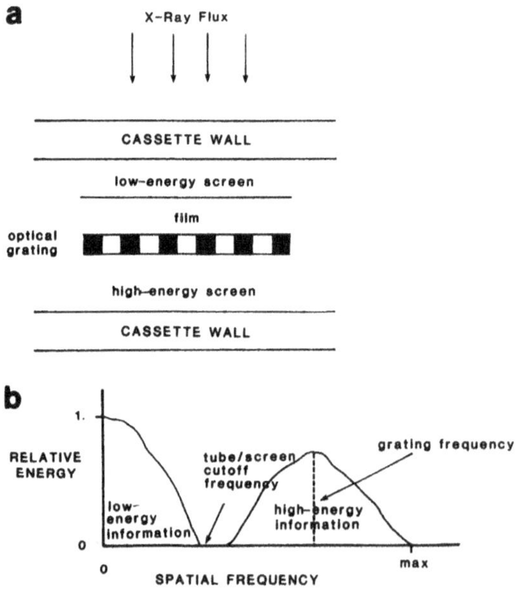

Figure 2. (a) Construction of a dual-energy single-film cassette. (b) Utilization of the film's full modulation transfer function for multiplexing dual-energy information.

wasted. One difficulty with this method is that it is not able to use an intermediate X-ray filter.

3. DECOMPOSITION OF THE ATTENUATION COEFFICIENT

In energy-selective radiography, it is important to understand the energy dependencies of X-ray interaction with matter. The true value of any model for imaging rests on: (1) the model's accuracy, relative to the accuracy of X-ray attenuation measurements, and (2) the mathematical consequences of the model's form. Although there are important nonlinearities in a detailed description of X-ray attenuation, linear system theory continues to be a very helpful tool. The concepts of modulation transfer function and reconstruction from line integrals have proven invaluable; the many useful results presented in the next sections illustrate the striking value of the linear approach. Any technique or model for extracting energy-dependent information must be linear if it is to utilize this powerful systems foundation.

Many nonlinear models of attenuation have been proposed for energy-selective imaging. These parameterization methods[25–27] attempt to extract information by using a parametric model of the linear attenuation coefficient as a function of atomic number, electron density, and X-ray energy. The approach has several problems. It is difficult to derive sufficiently accurate parametric expressions for the range of atomic numbers and energies of interest. But most fundamentally, it is difficult to rigorously generalize these expressions to realistic cases. The parameters, such as effective atomic number, are inherently nonlinearizable in the measured data. Because of this problem, a mixture of several elements requires an arbitrary definition of the effective parameters which cannot be justified on physical grounds.

A representative parameterization is of the form[27]

$$\mu(E) \sim \left\{ 9.8 \times 10^{-24} \frac{Z^{4.8}}{E^{3.2}} + 1.25 \times 10^{-24} \frac{Z^{3.0}}{E^{1.9}} + 0.0597 \times 10^{-24} e^{-0.0028\,(E-30)} \right\} n \quad (1)$$

and which includes terms modeling coherent scatter, incoherent scatter, and photoelectric absorption. Alvarez[11] proposed the following simple form which is free of any *a priori* constraint on the Z-dependent functions:

$$\mu(E) \sim \frac{a_1(Z)}{E^3} + a_2(Z)\sigma_{KN}(E) \quad (2)$$

This model led to two observations:

1. It has higher accuracy than the model of McCullough and Rutherford, owing to the freedom of the functional forms $a_1(Z)$ and $a_2(Z)$. The model error is of the same order as radiographic measurement uncertainty.
2. The model suggests a vector space representation of materials in which the functions E^{-3} and $\sigma_{KN}(E)$ are basis vectors.

This vector space representation has had a very strong impact on attenuation representations. In general, the vector space technique expands the attenuation coefficient as a linear combination of known functions of energy multiplied by undetermined coefficients:

$$\mu(E) = \sum_{i=1}^{N} a_i f_i(E) \tag{3}$$

Since the coefficients are the only unknown values in the expression, they carry all the information which can be extracted by an energy-selective imaging system. The representation is linear. It readily generalizes to a spatially varying form,

$$\mu(x,y,z,E) = \sum_{i=1}^{N} a_i(x,y,z) f_i(E) \tag{4}$$

and to a mixture of materials, where the total attenuation is the sum of component attenuations:

$$\mu(E) = \sum_{i=1}^{N} \sum_{j=1}^{M} \frac{\rho}{\rho_j} W_j a_{ij} f_i(E) \tag{5}$$

The choice of the basis functions $f_i(E)$ is crucial to this technique. Since the attenuation as a function of energy has only a countable number of discontinuities, a suitable set can be found. However, a set containing a very large number of functions (a large dimensionality) would be unsuited for practical applications. Fortunately, the minimum number of required functions is quite small.

The dimensionality of the vector space is an important property of the physical quantity being imaged. It determines the complexity of the procedure for extracting the energy-dependent information. What physical constraints limit the dimensionality of this vector space?

3.1. Intuitive Limits to Dimensionality

There are several important physical properties which must be incorporated into any description of the X-ray attenuation coefficient. One of these, the "mixture rule," states that the X-ray cross section of an atom is independent of its chemical state and that attenuation adds in proportion to the effective path length through each component material. This, of course, is not precisely true. A whole field of X-ray spectroscopy[(28)] is concerned with extracting information about chemical properties from these changes. However, these effects are confined to within a few kiloelectron volts of an absorption edge. Since most biological materials have their absorption edges at energies substantially below the diagnostic range, this effect can be ignored.

The mixture rule limits the dimensionality of the attenuation vector space to be less than or equal to the number of distinct elements within the object. For biological materials, this is a relatively small number. Of special interest is the observation that since the attenuation coefficient of any material is a linear combination of the attenuation coefficients of its constituent elements, it is itself a suitable candidate for a basis function.

That the dimensionality is further limited is suggested by another important physical property of X-ray attenuation, the "sum rule." Since X-ray photons are considered to be discrete, the various interactions are independent and mutually exclusive, and the total cross section is the sum of the cross sections for each type of interaction. Within the diagnostic region of energies, three types of interactions predominate: Compton scattering, photoelectric absorption, and Rayleigh (coherent) scattering. If these three cross sections can be expressed as separable terms, each a product of a function of atomic number only and another of energy only, then the dimensionality could be smaller than even the total number of elements. That is, if

$$\sigma(E) = a_{\mathrm{C}}(Z)f_{\mathrm{C}}(E) + a_{\mathrm{p}}(Z)f_{\mathrm{p}}(E) + A_{\mathrm{R}}(Z)f_{\mathrm{R}}(E) + \cdots \tag{6}$$

then the vector space dimensionality is equal to the number of different types of interactions. This, then, is the physical basis for Eq. (2), which includes Compton scattering and photoelectric absorption forms of interaction. This two-dimensional vector space representation of the attenuation coefficient has been found to be accurate to better than 0.5% fractional rms error for atomic numbers 1 to 25 and energies from 40 to 100 keV.

It is important to recognize that the vector space dimensionality is dependent on the measurement accuracy of the X-ray apparatus. Any representation of the attenuation coefficient is a simplified model and has an error. If that error is small compared to measurement inaccuracies, then it is likely that a simpler model with a smaller dimensionality will do just as well. A mathematical tool for quantifying this observation is described in the next section.

3.2. The Singular Value Decomposition

Alvarez[(29)] applied the singular value decomposition to quantify the dimensionality of the attenuation coefficient vector space. Singular value decomposition, a tool from matrix theory, helps to determine the rank of a matrix whose elements are known to contain errors.

Consider a matrix whose columns are the values of the attenuation coefficients of various elements at many diagnostic energies. If the elements found in biological materials are included, then by the mixture rule the attenuation coefficient of any biological material can be expressed as a linear combination of the columns of the matrix.

Analogous to the preceding discussion, if the entries in the matrix are considered to be of infinite precision, almost all matrices will have full-rank (i.e., the rank will be equal to the smaller of the number of rows or columns). However, in any physical situation, the matrix entries (measurements of the attenuation coefficient) will have limited accuracy. In this case, another matrix can be found which is "close" to the original matrix. If the "distance" between the two matrices is smaller than some

TABLE 1. Error in Representation of Attenuation as Dimensionality Increases, for Biological Materials and Diagnostic X-Ray Energies[a]

n	Normalized error
0	1.0
1	0.318
2	0.0037
3	0.0009
4	0.0005
5	0.0004
6	0.0003
7	0.0002

[a] A two-dimensional model is well matched to the accuracy of diagnostic X-ray measurements.

distance determined by measurement accuracies, then the two matrices have effectively the same rank.

Table 1 summarizes Alvarez's results for elements commonly found within biological materials ($Z = 1, 6, 7, 8, 15, 16, 17, 19, 20$) and for a range of energies (30–150 keV) used in medical radiographic applications. Calculated errors are expressed as a fraction of the attenuation coefficient. Note that there is always some reduction in error by using more functions in the basis set, but that the major reduction comes with two functions. The two-function basis set has an average error of 0.4%; this is less than or comparable to the error introduced by X-ray quantum noise in CT.

Figures 3–5 show Alvarez's results of attenuation coefficient error as more elements are considered and as the energy range is extended. Figure 3 shows the effect of an expansion of the range of elements included. As expected, errors increase somewhat with a larger range, but the increase in error, and, therefore, dimensionality, is not dramatic. Likewise, Figure 4 shows the effect of a wider range of energies. The high-energy limit is fixed at 500 keV, while the lower limit is permitted to vary. Figure 5 shows results for a set of elements including iodine for an energy range from 30 to 150 keV. Because iodine has an absorption discontinuity at 33.155 keV, the dimensionality of the space is now equal to three for even moderate accuracies.

The foregoing discussion should convince the reader that for biological materials within the diagnostic range of energies, a two-dimensional linear vector space representation of the linear attenuation coefficient

$$\mu(E) = a_1 f_1(E) + a_2 f_2(E) \tag{7}$$

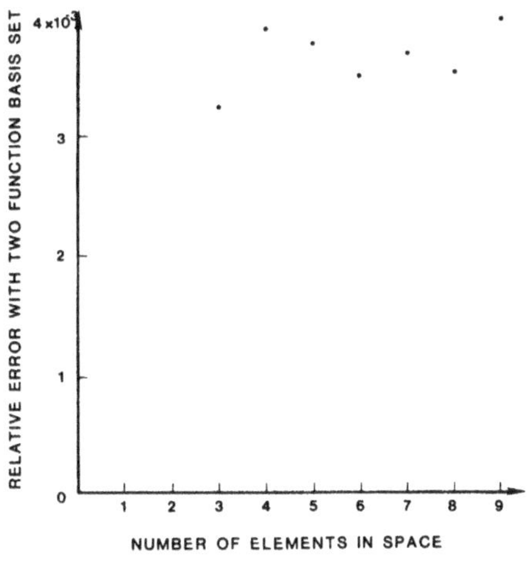

Figure 3. Relative mean squared error for a two-function basis set versus the number of elements in the space.

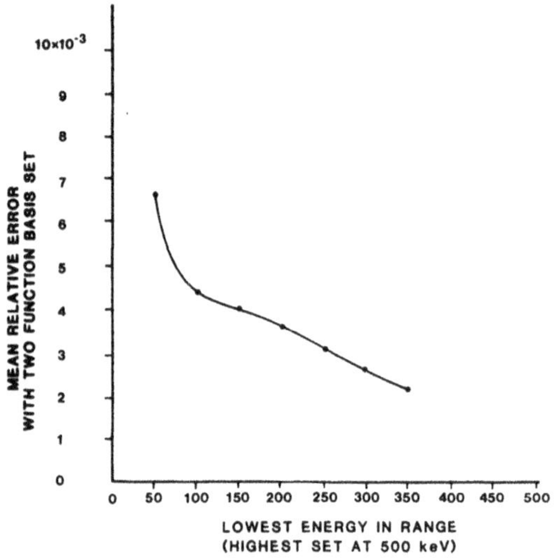

Figure 4. Relative mean squared error for a two-function basis set versus the energy range. The energy range is from the plotted value to 500 keV.

is sufficiently accurate even for the highly quantitative demands of CT. The "basis functions" $f_1(E)$ and $f_2(E)$ are fixed, and the "basis coefficients" a_1 and a_2 (or equivalently, the "basis vector" $\mathbf{a}$) represent all the energy-selective information that can be inferred from X-ray measurements of an elemental volume within the bounds of diagnostic accuracy.

One result of the singular value decomposition[29] is the set of optimal basis functions $f_i(E)$. Of course, that set is not unique and any nonsingular linear combination of those functions spans the same space. To the extent to which Eq. (7) is accepted as an equality, any two materials having

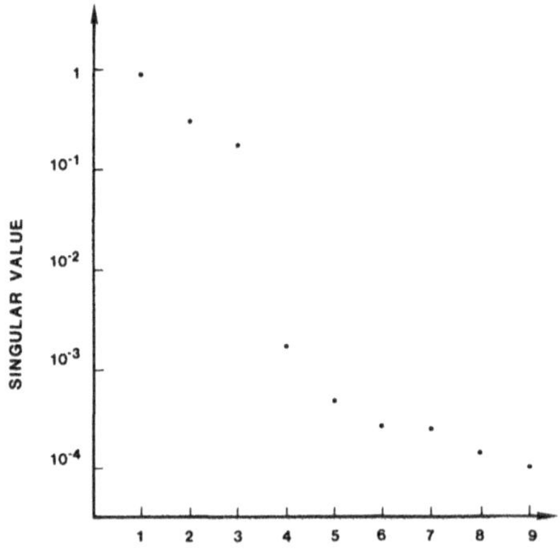

Figure 5. Plot of the singular values for a set of elements including iodine for energies from 30 to 150 keV. Since this range includes the iodine k-edge, more than two singular values are large and the basis set to represent the attenuation coefficients accurately must include more than two functions.

nonidentical basis vectors have attenuation coefficient functions of energy which are satisfactory basis functions.

4. CONDITIONS FOR CALCULATING COMPLETE ENERGY-DEPENDENT INFORMATION

4.1. Vector Space Descriptions of Mixtures and Line Integrals

A linear vector space representation uses simple geometric models to describe the attenuation of mixtures and line integrals. These models consist of a set of rules for manipulating coefficients directly without regard for the energy functions.

By the mixture rule, the total linear attenuation coefficient is related to the attenuation coefficients of the constituents by

$$\frac{\mu}{\rho} = \sum_{i=1}^{N} W_i \frac{\mu_i}{\rho_i} \tag{8}$$

where W_i is the function by weight of the element i in the mixture and ρ_i is its density.

Thus, if the component basis set coefficients are (a_{1i}, a_{2i}), then the basis set coefficients for the mixture are

$$a_1 = \sum_{i=1}^{N} W_i \frac{\rho}{\rho_i} a_{1i} \tag{9a}$$

$$a_2 = \sum_{i=1}^{N} W_i \frac{\rho}{\rho_i} a_{2i} \tag{9b}$$

The mixture's basis coefficients can be understood to be a weighted vector sum and given a geometric interpretation as shown in Figure 6.

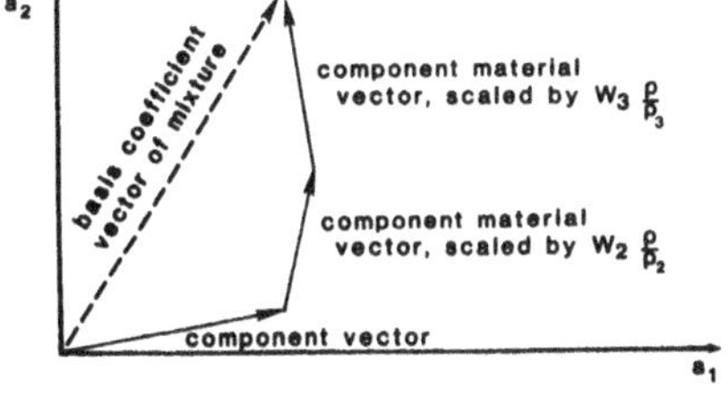

Figure 6. Geometric view that a mixture's basis coefficient vector is a vector sum of the basis vectors of the component materials.

A similar geometric interpretation holds true for the line integrals of a mixture:

$$L(E) = \int \mu(x,y,z,E)ds \tag{10}$$

Substituting the attenuation coefficient's linear basis decomposition,

$$L(E) = f_1(E) \int a_1(x,y,z)ds + f_2(E) \int a_2(x,y,z)ds \tag{11}$$

Denoting the line integrals of the basis coefficients A_1 and A_2, the line integral is then

$$L(E) = A_1 f_1(E) + A_2 f_2(E) \tag{12}$$

If the line integral path cuts through several materials with coefficients (a_{1i}, a_{2i}) and thicknesses l_i, then the line integrals can be more simply expressed:

$$A_1 = \sum_{i=1}^{M} a_{1i} l_i \tag{13a}$$

$$A_2 = \sum_{i=1}^{M} a_{2i} l_i \tag{13b}$$

Geometrically, these results describe the total line integral vector as a weighted vector sum of the component line integral vectors. This is illustrated in Figure 7.

4.2. Calculation of Line Integrals in Conventional Radiographic Systems

X-ray imaging systems do not measure the attenuation coefficient directly. Rather, they measure the flux transmitted through the object.

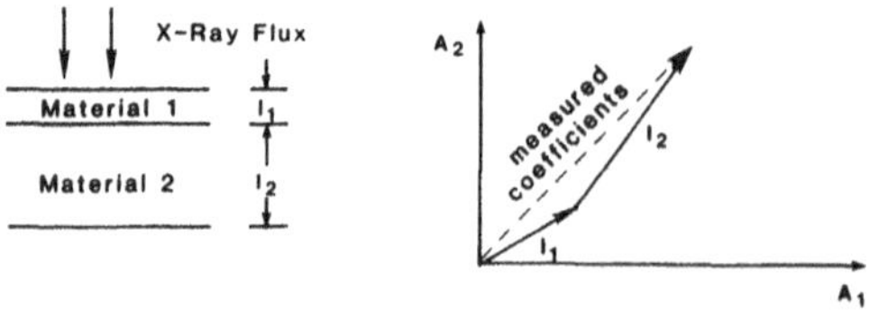

Figure 7. The line integral basis coefficient vector is the vector sum of the line integral vectors through component materials.

Neglecting scatter, and assuming an infinitesimally small beam, the transmitted flux is

$$I = \int S(E)e^{-L(E)}dE \qquad (14)$$

where $S(E)$ is the energy spectral density of the source and $L(E)$ is the line integral of the linear attenuation coefficient. In many quantitative applications, such as CT and digital radiography, it is desirable to calculate the line integral at a single energy from one flux measurement. Even if the spectrum is known, in general this is not possible.

Alvarez[29] proved the following result: the line integral at a single energy can be calculated from one flux measurement only if the spectrum is monoenergetic or if the object is known *a priori* to consist of a single material with known basis set coefficients.

4.3. Complete Information Extraction in Energy-Selective Systems

The preceding section demonstrated that, in general, the line integral cannot be calculated from a single flux measurement. This polyenergetic nonlinearity is responsible for artifacts in CT and for errors in quantitative digital radiography. Under quite general conditions, complete energy-dependent X-ray attenuation information (i.e., the line integral of the attenuation coefficient, at any energy within the diagnostic region) can be calculated from two flux measurements with different source spectra.

Equation (14) defines the transmitted flux as a function of two independent qualities, A_1 and A_2 (the source spectrum is assumed to be constant). Since there are two variables, the previous section demonstrated that, in general, this single measurement cannot be inverted to yield the line integrals. However, suppose another measurement is made with a different source spectrum. Then there are two functional relationships

$$I_1(A_1,A_2) = \int S_1(E) \exp\left[-A_1 f_1(E) - A_2 f_2(E)\right] dE \qquad (15a)$$

$$I_2(A_1,A_2) = \int S_2(E) \exp\left[-A_1 f_1(E) - A_2 f_2(E)\right] dE \qquad (15b)$$

which define a transformation between the measured quantities (I_1, I_2) and the desired results (A_1, A_2). Under what conditions is this transformation invertible?

Before discussing the situation in general, two special cases are considered. The first case is the use of two monoenergetic spectra with energies E_1 and E_2. The flux measurements can be linearized by taking the logarithm of the measurements. Then the logarithms and the line integrals are related by the linear equations

$$V_1 = \ln (I_1/I_{01}) = -A_1 f_1(E_1) - A_2 f_2(E_1) \quad (16a)$$

$$V_2 = \ln (I_2/I_{02}) = -A_1 f_1(E_2) - A_2 f_2(E_2) \quad (16b)$$

where I_{01} and I_{02} are the measured flux with zero object thickness. These equations can be uniquely solved for the line integrals if the determinant of the coefficients is not equal to zero:

$$\begin{vmatrix} f_1(E_1) & f_2(E_1) \\ f_1(E_2) & f_2(E_2) \end{vmatrix} \neq 0 \quad (17)$$

That is, if

$$\frac{f_1(E_1)}{f_2(E_1)} \neq \frac{f_1(E_2)}{f_2(E_2)} \quad (18)$$

Figure 8 shows a plot of the ratio of the basis set functions versus energy in the medical diagnostic region. Note that the ratio is monotonically decreasing. Thus, the condition of Eq. (18) will be true if the two energies are different.

A second special case is one in which the object is known to consist of a single material of thickness l. In this case, the line integrals are given by

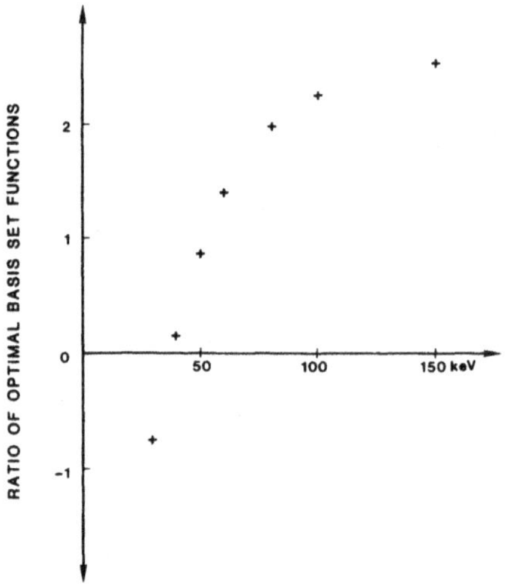

Figure 8. Plot of the ratio of basis set functions versus energy within the diagnostic region. The ratio is monotonically increasing.

$$A_1 = a_1 l \tag{19a}$$

$$A_2 = a_2 l \tag{19b}$$

The flux I is a function only of the thickness l,

$$I(A_1,A_2) = \int S(E) \exp\left\{-a_1 l\left[f_1(E) + \frac{a_2}{a_1} f_2(E)\right]\right\} dE \tag{20}$$

and the derivative of the flux

$$\frac{dI}{dl} = -\int [a_1 f_1(E) + a_2 f_2(E)]\, S(E) \exp \left\{-a_1 l\left[f_1(E) + \frac{a_2}{a_1} f_2(E)\right]\right\} dE \tag{21}$$

is always less than zero since the X-ray spectral density $S(E)$ and the attenuation coefficient are always positive. Thus, the flux is a monotonically decreasing function of the material thickness, and the flux can be used to calculate the thickness l (and therefore the line integral).

For the general case, Alvarez[29] shows that the values of the line integrals of the basis set coefficients can be calculated from transmitted flux measurements with incident spectra $S_1(E)$ and $S_2(E)$ with the following properties:

1. The maximum energies where $S_1(E)$ and $S_2(E)$ are nonzero are different for two spectra.
2. The Jacobian,

$$J = \begin{vmatrix} -\langle f_1\rangle_1 & -\langle f_2\rangle_1 \\ -\langle f_1\rangle_2 & -\langle f_2\rangle_2 \end{vmatrix} \tag{22}$$

   where $\langle\ \ \rangle_i$ represents an average over transmitted spectrum i, is nonzero over a region of the (A_1, A_2) plane containing the points for the object.

If these conditions are satisfied, then the two flux measurements can be used to give the line integrals of the attenuation coefficient at all energies within the diagnostic region. These line integrals are all the information which can be produced by a single projection system. In a CT system, the line integrals can be used to give the values of the basis set coefficients at points in the object's cross section.

5. APPLICATIONS OF ENERGY-SELECTIVE IMAGING

This section describes a variety of previously unobtainable forms of images which all derive from energy-selective data. The mathematical operations generating all these image types have a common form called the "generalized projection," a weighted sum of two "basis images" (A_1, A_2) or (a_1, a_2). The generalized projection is geometrically equivalent to the projection of the basis vector onto another given vector, the angle of which defines the types of images to be generated.

For notational simplicity, consider "a basis projection" at angle ϕ. This is a linear combination of the two basis images A_1 and A_2 with weights which are the following function of ϕ:

$$C = A_1 \cos \phi + A_2 \sin \phi \tag{23}$$

where C is the basis projection image. This section presents the deterministic properties of C as ϕ ranges from 0 to 2π radians.

5.1. Synthesized Monoenergetic Images

Perhaps the simplest form of processing is to form images representing the same physical quantities imaged in conventional systems, the linear attenuation coefficient $\mu(E_0)$ (in CT) or its line integral

$$L(E_0) = \int u(x,y,z,E_0)ds \tag{24}$$

in single projection systems. The data from energy-selective systems can be used to form images similar to those of conventional systems, but without beam-hardening artifacts and at average energies not determined by the physical characteristics of the system but by an arbitrary parameter under our control. These images actually do represent a single energy and they are called synthesized monoenergetic images.

The calculation of the synthesized monoenergetic image is based on the fundamental vector space representation

$$\mu(E) = a_1 f_1(E) + a_2 f_2(E) \tag{25}$$

In an energy-selective CT system, where a_1 and a_2 are determined at points in the object cross section, a display image can be calculated by carrying out the mathematical operation indicated at every point in the image. The operation is simply a weighted sum of two "basis images," where the relative weights are related in a predictable way to the arbitrary display energy. Since the basis functions $f_1(E)$ and $f_2(E)$ are known *a priori*, any desired energy can be used. Analogous results can be obtained for the line integral of the attenuation coefficient $L(E)$, which can be expressed in terms of the line integrals of the basis set coefficients A_1 and A_2 as

$$L(E) = A_1 f_1(E) + A_2 f_2(E) \qquad (26)$$

The synthesized monoenergetic image has a useful vector interpretation. The fundamental decomposition, Eq. (25), can be considered to be an inner product between two vectors. One, with components (a_1, a_2), depends on the material characteristics of the object, while the other, with components $[f_1(E_0), f_2(E_0)]$, depends only on the selected display energy. The operation of calculating a conventional image from the energy-selective information is then equivalent to projecting the vector representing the basis set coefficients onto a vector representing the values of the basis functions at the display energy and then adjusting the scale. Thus, the conventional image, neglecting noncorrectable beam-hardening nonlinearities, is a subset of the possible images obtainable from energy-selective data. An energy-selective system always has the possibility of extracting more information than a conventional system.

The utility of synthesized monoenergetic images depends on the application. They are quite important in CT since they are images which have been corrected for beam-hardening artifacts.[30] These artifacts can have serious effects in quantitative applications.

5.2. Selective Material Images

Unlike monoenergetic images, another group of energy-selective images called "selective material images" are distinct from the familiar conventional images. No possible source spectrum can generate images like these. These images have the unusual property that one type of material can have an effective attenuation coefficient of zero. It has no contrast and apparently vanishes from the image. The apparent attenuation coefficients of other materials change as well. Some appear positive and others appear negative.

What is the value of selective material images? Those familiar with radar speak of power-limited and clutter-limited cases. In the power-limited case, image quality is determined mostly by random noise; quality improves with increasing transmitter power (corresponding to an increased radiographic exposure). In the clutter-limited case, image quality is limited by spurious and undesired but deterministic signals and image quality does not improve with greater transmitted power. This case calls for ingenuity and a novel approach to data acquisition and signal processing.

This latter case, where anatomical shadows or "clutter" limit the visibility of other structures, is often a good model of single-projection radiography. In radiography, the term "conspicuity" is used to mean the degree to which a selected feature stands out from the background clutter. Conspicuity is intimately related to the workings of the observer's psychovisual process, a system which is so awesomely complex as to preclude quantification and analysis. Still, with all geometries held fixed, it is reasonable to expect conspicuity to vary directly with the contrast of the selected feature relative to the background.

Selective material imaging relies on the observation that the vector representation for a given material always lies along a single line. The line integrals of the basis coefficients of a simple homogeneous material are

$$A_1 = a_1 l \tag{27a}$$

$$A_2 = a_2 l \tag{27b}$$

where l is the thickness of the material along the X-ray beam path. As l is permitted to vary, the line integral vectors (A_1, A_2) fall on a straight line through the origin with slope s,

$$s = a_2/a_1 \tag{28}$$

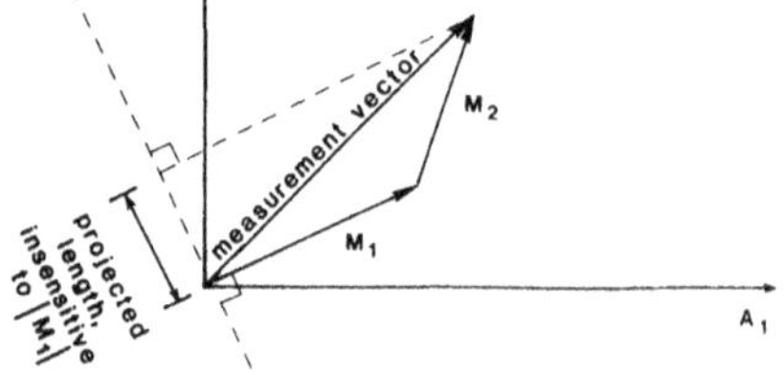

Figure 9. Vector interpretation of selective material imaging. A point-by-point projection of the measurement vector (a vector sum of component material vectors) onto a line orthogonal to the material M_1, will be insensitive to the presence of M_1.

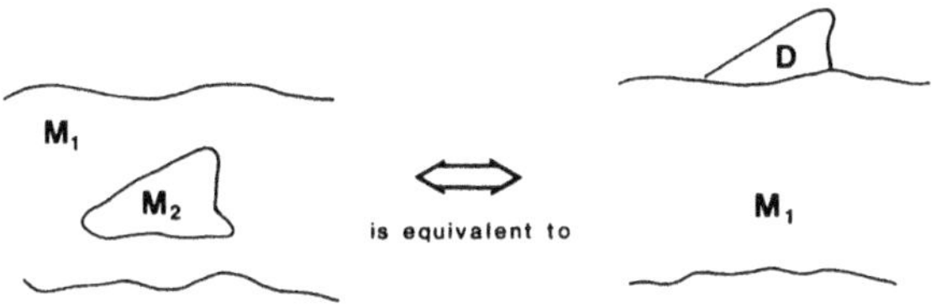

Figure 10. A feature (M_2) surrounded by, and displacing, the background material (M_i). To remove feature contrast is equivalent to removing a virtual material which can be represented by the vector difference between feature and background.

If a generalized projection is formed point by point through an image perpendicular to this line, then variations in the material's thickness will not contribute to the image (Figure 9).

Suppose that in a single-projection system the object consists of a feature of interest plus a constant background. Then the total vector equals the vector sum of a background vector $\mathbf{A}_b$ plus a feature vector $\mathbf{A}_f$. Making a generalized projection at an angle ϕ_b perpendicular to $\mathbf{A}_b$ will cancel the background material while a generalized projection at an angle ϕ_f perpendicular to $\mathbf{A}_f$ will cancel the feature.

The cancellation can easily be described from a nongeometric interpretation. For the object described above, the line integrals of the basis set coefficients will be

$$A_1 = a_{b_1} l_b + a_{f_1} l_f \tag{29a}$$

$$A_2 = a_{b_2} l_b + a_{f_2} l_f \tag{29b}$$

Multiplying the first equation by a_{b_2} and the second by $-a_{b_1}$ and adding the results gives

$$A_1 a_{b_2} - A_2 a_{b_1} = (a_{b_1} a_{b_2} - a_{b_2} a_{b_1}) l_b + (a_{f_1} a_{b_2} - a_{f_2} a_{b_1}) l_f \tag{30}$$

Since the first term is zero, the result depends only on l_f, the amount of the feature material. If this operation is carried out at every point in the image, the thickness of the background material will not affect the result. It will have been canceled from the image. Note that this result is true regardless of whether the background material is constant in thickness. Note also that given a basis space dimensionality of two, at most one specific material can be removed at one time from an image. One specific material is defined as the set of all those substances whose basis vectors (a_1, a_2) have the same characteristic angle θ:

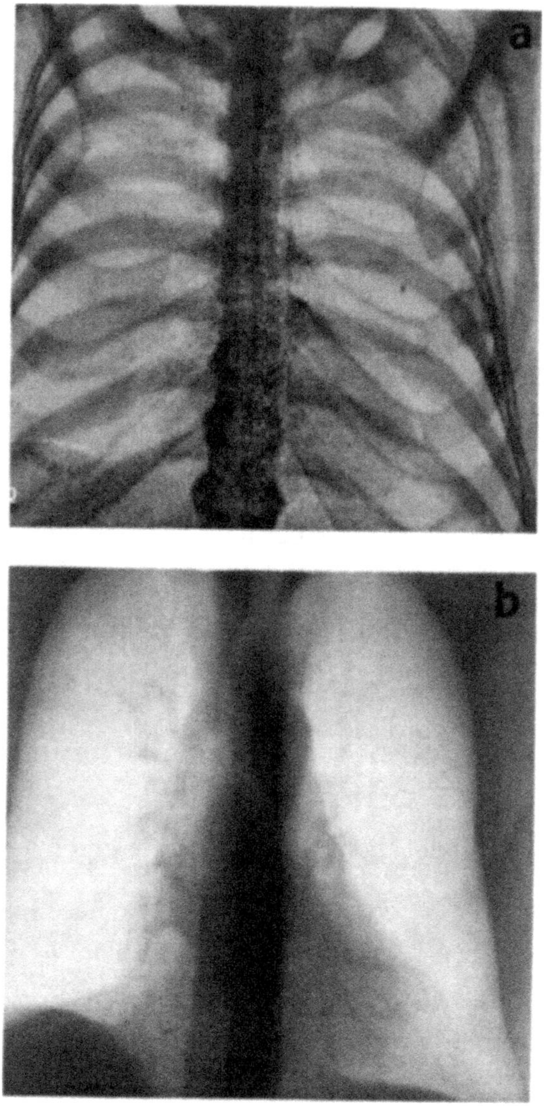

Figure 11. Clinical examples of selective material imaging. Part (a) shows a soft-tissue-canceled chest image. In (b) the bone has been suppressed. Parts (c) and (d) show conventional and soft-tissue-canceled intravenous pyelograms.

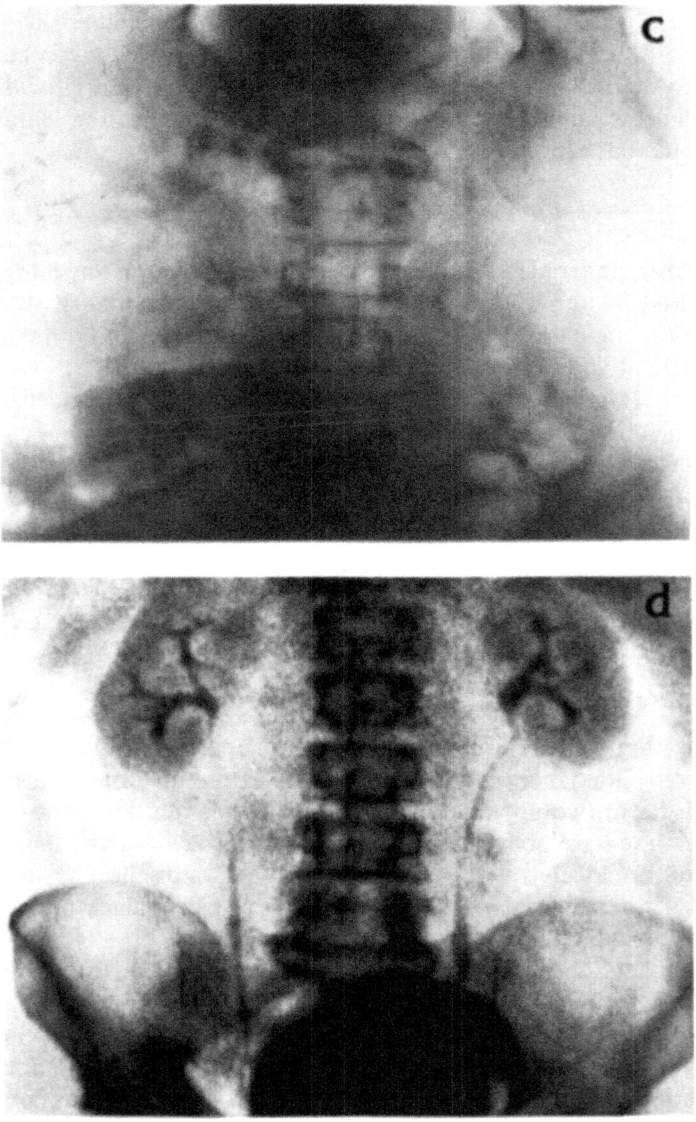

Figure 11. (*continued*)

$$\tan \theta = a_2/a_1 \tag{31}$$

From X-ray measurements these substances cannot be distinguished.

If the object consists of a feature of interest completely surrounded by a uniform background, the results are similar. This case can be transformed to the previous case by defining an effective material with basis set coefficients equal to the difference between those of the feature and those of the background (Figure 10):

$$\mathbf{a}_d = \mathbf{a}_f - \mathbf{a}_b \tag{32}$$

By forming a generalized projection perpendicular to the line integrals of the effective material $\mathbf{a}_d$, the embedded material will vanish. Now the cancellation angle depends on the basis set coefficients of both materials and their relative densities.

Figure 11 gives some clinical examples of material cancellation in single projection. Parts (a) and (b) show soft-tissue- and bone-canceled chest images. Removing the bone greatly increases the visibility of soft tissue structures underlying the bone. Parts (c) and (d) show conventional and soft-tissue-canceled images of the lower abdomen in an intravenous pyelogram. Here, overlying soft tissue and bowel gas reduce the visibility of the contrast material. Removing the contribution of the soft tissue aids in contrast agent visibility.

5.3. Generalized Projection Signal Processing

Thus far two uses of the basis coefficients have been discussed: synthesized monoenergetic images, and selective material cancellation. Although these two applications image fundamentally different physical quantities, they are both linear combinations of the two basis images and are therefore closely related. The following explores the physical roots of this relation and shows that no other types of images of the same general type exist.

Consider a basis representation of the form of Eq. (6) with photoelectric absorption $f_1(E)$ and Compton scatter $f_2(E)$ as the two basis functions of energy. For this basis set, the coefficients are always positive and thus the material vectors are always found in the first quadrant. All selective material cancellation projection vectors are orthogonal to the material vectors and therefore lie in the second or fourth quadrants:

$$\phi = \tan^{-1}(-a_2/a_1) \tag{33}$$

It can be shown[31] that the ratio of basis set coefficients is a monotonically decreasing function of the atomic number Z. Thus, there is a one-to-one mapping of projection angle ϕ onto atomic number of the canceled substance.

Likewise, in forming the synthesized monoenergetic generalized projection,

$$c = a_1 f_1(E_0) + a_2 f_2(E_0) \tag{34}$$

the projection vector $[f_1(E_0), f_2(E_0)]$ lies in the first quadrant. As noted earlier, the ratio of the basis functions is a monotonically decreasing function of energy, and there is a one-to-one mapping of the effective display energy E_0 into the projection angle ϕ:

$$\phi = \tan^{-1} [f_2(E_0)/f_1(E_0)] \tag{35}$$

There are two disjoint regions of projection angle. The first quadrant corresponds to synthesized monoenergetic images, and the second quadrant corresponds to selective material images. Projection vectors within the third and fourth quadrants generate images which are contrast reversed from those in the first and second quadrants, respectively, but which contain no new information. For basis functions other than the photoelectric/Compton set just discussed, the two regions remain disjoint but rotate through an angle with the basis transformation. Figure 12 depicts the mapping of projection angle into display energy and atomic number for an aluminum and Lucite basis set.

5.4. Computation for Energy-Selective Imaging

Recall, from Eq. (14), that the flux measurements are related to the basis coefficients (A_1, A_2) as

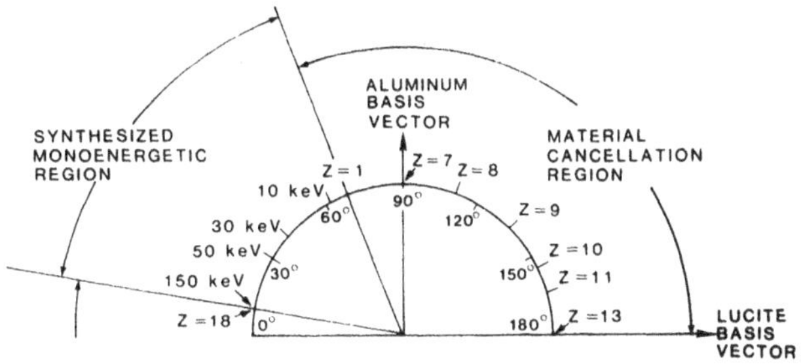

Figure 12. The mapping of projection angle into display energy and atomic number of canceled material. The two regions are disjoint and preclude the presence of any other regions.

$$I = \int S(E) \exp\left[-A_1 f_1(E) - A_2 f_2(E)\right] dE \qquad (36)$$

Following the acquisition of energy-selective flux measurements, the I_i, it is necessary to invert this nonlinear integral equation to solve for the basis coefficients A_i. When the process is repeated for every point in the projection image, the result is a pair of basis images representing A_1 and A_2 along every projection path.

In the typical situation, $S(E)$ is not known from first principles, depending as it does on the X-ray tube, its operating conditions, inherent filtration, and the response characteristics of the detector. Thus, the approach provides for self-calibration.

When the $S_i(E)$ are monoenergetic, at energies E_1 and E_2, then there is a simple relationship consisting of two linear simultaneous equations and four yet undetermined coefficients:

$$V_1 = \ln(I_1/I_{01}) = -f_1(E_1)A_1 - f_2(E_1)A_2 \qquad (37a)$$

$$V_2 = \ln(I_2/I_{02}) = -f_1(E_2)A_1 - f_2(E_2)A_2 \qquad (37b)$$

When the conditions specified in Section 4 for the extraction of complete energy-selective information are met, this relationship is invertible

$$A_1 = k_1 V_1 + k_2 V_2 \qquad (38a)$$

$$A_2 = k_3 V_1 + k_4 V_2 \qquad (38b)$$

where, again, the V_i are the logarithms of the measurements:

$$V_i = \ln(I_i/I_{0i}) \qquad (39)$$

For convenience in calibration, we choose for basis functions (see arguments at the conclusion of Section 3) the attenuation coefficient functions of energy of two common materials which span the range of atomic numbers encountered in biological objects. Aluminum and lucite are most frequently chosen. They are easily shaped and formed to precise tolerances. The atomic number of aluminum (13) is higher than that of bone (12.7), and the effective atomic number[(31)] of lucite (6.6) is lower than that of water.

Typically, precise overlapping wedges of aluminum and lucite are constructed and imaged so that for each thickness of the one material there are many overlapping steps of the other. The Stanford University wedge set[(31)] provides 88 thickness combinations. The constants k_i are then evaluated in a least-squares sense during the calibration procedure in which the true basis coefficients (A_1, A_2) of each wedge step are known in advance. The estimation errors may be weighted inversely with measurement noise. This "weighted fit" shifts the deterministic fit errors to be smaller for those steps where measurements are more accurate:

$$k_i;\ i = 1,n \text{ minimize} \sum_m \frac{(A_m - \hat{A}_m)^2}{\sigma_{I_1}^2 + \sigma_{I_2}^2} \tag{40}$$

Now real sources introduce polyenergetic nonlinearities. With typical sources, the monoenergetic assumption of Eq. (37) gives an error of approximately 6% rms over a common range of absorptive materials. A higher-order approximation is better able to fit the nonlinear character of Eq. (36) over a wide range(32):

$$\hat{A}_1 = k_1 V_1 + k_2 V_2 + k_3 V_1^2 + k_4 V_2^2 + k_5 V_1 V_2 \tag{41a}$$

$$\hat{A}_2 = k_6 V_1 + k_7 V_2 + k_8 V_1^2 + k_9 V_2^2 + k_{10} V_1 V_2 \tag{41b}$$

The general polynomial can be of any order and Eq. (40) remains a valid method for finding the undetermined coefficients.

Reflecting to a greater extent the physics of attenuation, an alternative "iterative" form of computation was suggested by Alvarez:

$$V_1 = k_1 \hat{A}_1 + k_2 \hat{A}_2 + k_3 \hat{A}_1^2 + k_4 \hat{A}_2^2 + k_5 \hat{A}_1 \hat{A}_2 \tag{42a}$$

$$V_2 = k_6 \hat{A}_1 + k_7 \hat{A}_2 + k_8 \hat{A}_1^2 + k_9 \hat{A}_2^2 + k_{10} \hat{A}_1 \hat{A}_2 \tag{42b}$$

These simultaneous equations are solved iteratively using the Newton–Raphson method. The linear terms of a closed-form solution act as a good first guess and the equations converge in fewer than three interactions on the average. The iterative method forms a better fit for the same order than does the closed form expression given previously, but at a greater cost in computation.(32) Figure 13 compares weighted and unweighted closed-form approximations of order one and two for typical spectra.

The determination of basis images is just an interim step. All nonlinearities have been removed; the basis image data are quantifiable and system independent. CT Hounsfield units, by way of contrast, are not system independent since they depend on the spectral properties of the CT X-ray system.

But the basis images exist in three dimensions: two of space and one of energy. This complicates display somewhat and a simple and straight-

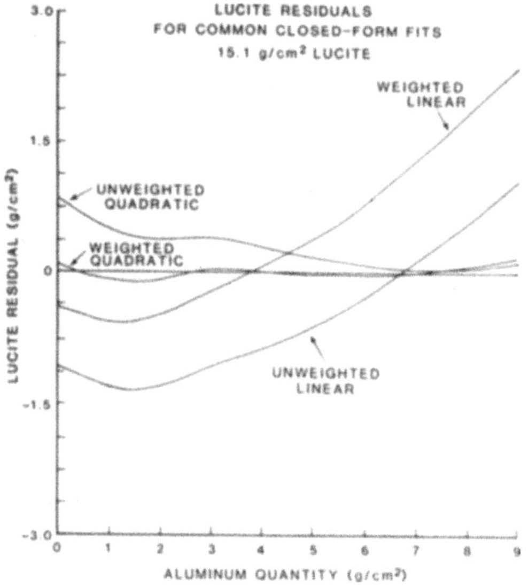

Figure 13. A comparison of weighted and unweighted closed-form polynomial approximations to the inversion of the imaging function. For typical spectra and objects, a second-order polynomial has a fit error of no more than 1%.

forward display of the basis images themselves is not possible. The concept of generalized projections (Section 5.3) has been introduced as a way of reducing the inherent dimensionality of dual-energy imaging to two.(33) Any particular basis projection image, however, displays only a subset of the acquired data. That is why the following argument is made for a dynamic and interactive display in which time is used to add an extra dimension to the display.

The human observer is an essential part of the dual-energy radiography system. The computation elements of the system adjust the relative contrasts of the object's features; the observer performs the ultimate feature recognition and evaluation. Feature recognition, a central element of all imaging tasks, has particular importance in feature identification by iterative cancellation.

At times, a set of standard basis projections suffice to provide satisfactory images. The water-canceled image and the bone/water look-alike image are sufficient in many cases. At other times man and machine must interact to arrive, finally, at an optimal image. There are three circumstances which require iterative display:

1. When identifying an unknown. The observer adjusts the basis projection angle to cancel a selected feature. The observer notes the angle of cancellation to characterize the feature.
2. When imaging a complex object composed of multiple desired and

undesired features. The optimal contrast ratios and signal-to-noise ratios (which also depend on the basis projection angle) are usually not calculable, as they depend on the observer's psychovisual process and the geometry of features.

3. When the optimal parameters are spatially variant. The set of optimal contrast ratios (and hence the set of optimal projections) frequently vary over the image area. Each region demands locally optimal processing, just as window and level settings may be locally optimal.

A sequence of fixed images is probably not as useful as an adjustment by which the operator can "follow" a feature. The observer works most efficiently when the iteration cycle time is on the order of tens of milliseconds. At these rates the native ability to measure change and rates of change is sharpest. For this reason, level and window computations are often performed at video rates. Experience at Stanford University's energy-selective digital radiography system has shown the value of real-time operator control over the basis projection parameter.

6. ANALYSIS OF CONSPICUITY AND NOISE

6.1. Statistics of Basis Coefficient Estimation

Alvarez,[11] and later Rutt and Fenster,[34] expanded the inverse of the imaging equation (14) in a Taylor series about the two log measurements $\overline{V}_1$ and $\overline{V}_2$. Consider two functions g_1 and g_2 for which

$$\begin{aligned} \hat{A}_1 &= g_1(V_1, V_2) \\ \hat{A}_2 &= g_2(V_1, V_2) \end{aligned} \tag{43}$$

Then

$$\begin{aligned} \hat{A}_1 = {} & g_1(\overline{V}_1, \overline{V}_2) + \frac{\partial g_i}{\partial V_1}(V_1 - \overline{V}_1) + \frac{\partial g_i}{\partial V_2}(V_2 - \overline{V}_2) \\ & + \frac{1}{2}\frac{\partial^2 g_i}{\partial V_1^2}(V_1 - \overline{V}_1)^2 + \frac{1}{2}\frac{\partial^2 g_i}{\partial V_2^2}(V_2 - \overline{V}_2)^2 + \cdots \qquad (i = 1, 2) \end{aligned} \tag{44}$$

Now the polyenergetic nonlinearities in the g_i are very modest on an absolute scale. Further, for the high photon rates typical of diagnostic imaging, the measurement noise is relatively small. Over the small interval

ΔV_1, ΔV_2, the g_i is essentially linear. Alvarez and Rutt truncated the series after the linear terms:

$$\hat{A}_i \sim g_i(\overline{V}_1,\overline{V}_2) + \frac{\partial g_i}{\partial V_1}(V_1 - \overline{V}_1) + \frac{\partial g_i}{\partial V_2}(V_2 - \overline{V}_2) \tag{45}$$

The estimate bias is insignificant,

$$\text{BIAS} = \overline{\hat{A}_1 - A_1} \sim \hat{A}_i - g_i(\overline{V}_1,\overline{V}_2) = 0 \tag{46}$$

and the estimate variance is simply

$$\text{VARIANCE} = \overline{(\hat{A}_i - A_i)^2} \sim \frac{\partial g_i}{\partial V_1}\sigma^2_{V_1} + \frac{\partial g_1}{\partial V_2}\sigma^2_{V_2} \tag{47}$$

Some researchers have speculated that weighted sums and differences of the measurements are in some sense better conditioned and lead to smaller estimate variances. The University of Toronto group[34,35] estimated photoelectric and Compton components from the sum and difference of their measurements. Perhaps they were motivated by the idea of a mathematical spectral filter for reducing the spectral overlap in the measurements. If a virtual measurement is formed by linearly combining physical measurements,

$$I'_j = \sum_{i=1}^{n} a_{i_j} I_i \tag{48}$$

the imaging integral becomes

$$\overline{I}'_j = \int \left[\sum_{i=1}^{n} a_{i_j} S_i(E)\right] \exp\left[-A_1 f_1(E) - A_2 f_2(E)\right] dE \tag{49}$$

Equation (49) suggests that a virtual spectrum has been formed,

$$S'(E) = \sum_{i=1}^{n} a_i S_i(E) \tag{50}$$

which may be preferred to physically realizable spectra. Lehmann[31] proved that the basis estimate variance is not affected by postacquisition manipulations. Image variance depends on the quality of the measurements themselves. Once the data are recorded, the variance is already determined for a given imaging task. These simple results follow directly from the linear, small-signal approximation: "Basis estimate variance is invariant under (nonsingular) linear combinations of measurements."[31]

The basis estimate variance continues to be independent of any smooth and nonlinear function of the measurements so long as the small-signal linear analysis remains valid. Of course, the number of virtual measurements must, in any case, equal the number of basis components.

The small-signal formalism above readily generalizes the earlier variance calculation (47) to dimensions greater than two and represents the same results in a more convenient form, as will be shown later. With matrix M defined by the relationship,

$$\mathbf{\Delta A} = M^{-1}\mathbf{\Delta I} \tag{51}$$

Then the system covariance matrix, $\mathbf{R}_A$, is

$$\mathbf{R}_A = E(\mathbf{\Delta A} \cdot \mathbf{\Delta A}^T) = \mathbf{M}^{-1}\mathbf{R}_I\mathbf{M}^{-T} \tag{52}$$

where $\mathbf{R}_I$ is the measurement covariance matrix, having zeros off the main diagonal,

$$\mathbf{R}_I = \begin{bmatrix} \sigma_{I_1} & & 0 \\ & \ddots & \\ 0 & & \sigma^2_{I_n} \end{bmatrix} \tag{53}$$

The variances of the basis coefficients are, of course, the diagonal elements of the system covariance matrix.

The variances and covariance of the basis line integrals can be expanded, for dimensions of two, in the form previously used by Alvarez:

$$\mathrm{VAR}\,(A_1) = \frac{m_{22}^2\sigma_{I_1}^2 + m_{12}^2\sigma_{I_2}^2}{D^2} \tag{54a}$$

$$\mathrm{VAR}\,(A_2) = -\frac{m_{21}^2\sigma_{I_1}^2 + m_{11}\sigma_{I_2}^2}{D^2} \tag{54b}$$

$$\mathrm{COV}\,(A_1, A_2) = \frac{-m_{22}m_{21}\sigma_{I_1}^2 - m_{11}m_{12}\sigma_{I_2}^2}{D^2} \tag{54c}$$

where

$$D = m_{11}m_{22} - m_{12}m_{21} \tag{55}$$

and $\sigma^2_{I_i}$ are the measurement variances.

6.2. Basis Noise and the System's Physical Properties

There are two factors which determine the overall uncertainty in the basis vector estimates (A_1, A_2). One is the fundamental noise of X-ray measurements due to the quantized nature of X-ray interactions with matter. In systems using counting detectors, the measurement variance is[36]

$$\text{VAR}\,[\log(N)] = 1/\overline{N} \tag{56}$$

under most circumstances. For energy-integrating detectors, which are more commonplace in diagnostic imaging, the measurement variance is[36]

$$\text{VAR}\,[\log(Q)] = \frac{1}{N}\frac{\overline{E^2}}{\overline{E}^2} \tag{57}$$

where N is the total number of photons detected, $\overline{E}$ is their average energy, and $\overline{E^2}$ is the average of the square of their energies. Equations (54) have in their numerators terms which represent the accuracies of the two measurements.

The other factor determining overall basis vector uncertainty is the conditioning of the linear equations (38). This is, in turn, determined by suitably defined "effective energies" of the measurement spectra. Each of the equations (54) has a factor in the denominator representing matrix conditioning. The M matrix which defines the value of these denominators has been shown by Alvarez[37] to be

$$m_{i_j} = \frac{\partial \log N_i}{\partial A_j} = \frac{1}{N_i}\frac{\partial N_i}{\partial A_j} \tag{58}$$

$$= -\frac{\int f_j(E) S_i(E) \exp\,[-\ A_1 f_1(E) + A_2 f_2(E)]dE}{\int S_i(E) \exp\,[-\ A_1 f_1(E) + A_2 f_2(E)]dE} \tag{59}$$

That is, m_{ij} is the negative of the average value of $f_j(E)$ over the effective spectrum detected in measurement i. In light of the theorem presented in Section 6.1, there is no advantage to one set of basis vectors over another within a given space. Therefore, the effective numerical conditioning is determined solely by the qualities of the measurement spectra. They should be minimally overlapping, emphasize the difference between photoelectric absorption and Compton scatter, and provide for high photon counts at the detector.

6.3. Noise Optimal Generalized Projections

In the previous sections, the noise properties of the basis vector estimates were analyzed. Although the orthogonal components of the basis vector are sometimes viewed, it is more common to view a weighted sum of the vector components, a basis projection image. It is here where there is most interest in noise.

As will be shown, noise in the projection image is a function not only of noise in the basis estimates but also of the basis projection angle ϕ. This parameter ϕ determines the imaging task as well as its noise properties. One value of ϕ will produce a synthesized monoenergetic image at, say, 92 keV, free of polyenergetic nonlinearities; another value of ϕ will produce an image in which water does not contribute, bone contributes normally, but in which fat has a *negative* contribution! It is important here to consider the noise properties of the parameter ϕ when the deterministic aspects of the set of images under consideration vary.

It makes sense to treat the synthesized monoenergetic and material-canceled types of images separately. They image fundamentally different quantities, and the task in the case of material-canceled images is usually well defined. When no *a priori* information is available, the synthesized monoenergetic type of image is usually most appropriate. As ϕ varies through this region, i.e., as the display energy varies, the image noise varies as well and there is an optimal display energy which gives a minimum variance. On the other hand, when some information is available in advance and the imaging task is better defined, the material-canceled image is the most suitable; in this case an SNR can be defined relevant to the task at hand and a value found for ϕ, the basis projection angle, which maximizes that measure.

Consider first the synthesized monoenergetic images and the projection which minimizes the noise variance. The variance of a linear combination

$$C = C_1A_1 + C_2A_2 \tag{60}$$

of the basis set line integrals is

$$\mathrm{VAR} = C_1^2\,\mathrm{VAR}\,(A_1) + C_2^2\,\mathrm{VAR}\,(A_2) + 2C_1C_2\,\mathrm{COV}\,(A_1,A_2) \quad (61)$$

Using the expressions derived previously, the noise variance of the projection image is then

$$\mathrm{VAR}(C) = \frac{(C_1m_{22} - C_2m_{21})^2\sigma_{\bar{I}_1}^2 + (C_1m_{12} - C_2m_{11})^2\sigma_{\bar{I}_2}^2}{D^2} \quad (62)$$

For a synthesized monoenergetic image, the combination coefficients are

$$C_1 = f_1(E_\mathrm{d}) \quad (63)$$

$$C_2 = f_2(E_\mathrm{d}) \quad (64)$$

For display energy E_d, Alvarez[37] shows that the optimal display energy, $E_{\mathrm{d}_\mathrm{opt}}$, is defined implicitly by

$$f_1(E_{\mathrm{d}_\mathrm{opt}}) = -\frac{m_{21}\sigma_{\bar{I}_1}^2 + m_{11}\sigma_{\bar{I}_2}^2}{\sigma_{\bar{I}_1}^2 + \sigma_{\bar{I}_2}^2} \quad (65\mathrm{a})$$

$$f_2(E_{\mathrm{d}_\mathrm{opt}}) = -\frac{m_{22}\sigma_{\bar{I}_1}^2 + m_{12}\sigma_{\bar{I}_2}^2}{\sigma_{\bar{I}_1}^2 + \sigma_{\bar{I}_2}^2} \quad (65\mathrm{b})$$

At the optimal display energy, the variance of the synthesized monoenergetic image is simply

$$\mathrm{VAR}\,(C) = \frac{\sigma_{\bar{I}_1}^2\,\sigma_{\bar{I}_2}^2}{\sigma_{\bar{I}_1}^2 + \sigma_{\bar{I}_2}^2} \quad (66)$$

For counting detectors, the variance

$$\mathrm{VAR}\,(C) = \frac{1}{N_1 + N_2} \quad (67)$$

is equal to the variance of a conventional image produced using the same number of detected photons.

The variance in Eq. (66) does not depend on the measurement spectra (except to the extent that their "average energies" be different) and hence the conditioning of the measurement process. This result is used in CT to produce a beam-hardening-corrected synthesized monoenergetic image from two spectra which are quite ill conditioned.[34] The noise in the optimal image will be as small as for a well-conditioned measurement set. However, Alvarez[37] has shown that the sharpness of the optimum becomes larger as the conditioning becomes worse. Since the transmitted spectrum depends on the object, if the measurement set is too ill conditioned then it may not be possible to define a single useful optimal energy for the whole object.

When certain details of the object and imaging task are known in advance, a consideration of the SNR is more useful than simply an interest in noise without regard to the scale of signal. For example, there is the simple situation of a feature of interest over a uniform background. If the compositions of feature and background are known, it is possible to predict the value of ϕ which maximizes the SNR, improving the confidence in distinguishing a feature from background. The optimal angle depends on the noise properties of the system as well as the composition of the object.

Define SNR as the projection through feature and background minus the projection through background alone, and normalized by the standard deviation

$$\mathrm{SNR} = \Delta C/\sigma C \tag{68}$$

where

$$\Delta C = |C_{\mathrm{b+f}} - C_{\mathrm{b}}| \tag{69}$$

Now the projection image is a weighted sum of two basic images, expressed in vector notation as

$$C = \mathbf{K}^T\mathbf{A}, \qquad \mathbf{K}^T = [k_1, k_2] \tag{70}$$

The variance of the scalar projection C is

$$\mathrm{VAR}(C) = \mathbf{K}^T R_A \mathbf{K} \tag{71}$$

and the SNR in C is

$$\mathrm{SNR} = \frac{\mathbf{K}^T(\mathbf{A}_{\mathrm{b+f}} - \mathbf{A}_{\mathrm{b}})}{[\mathbf{K}^T R_A \mathbf{K}]^{1/2}} \tag{72}$$

The projection that maximizes the SNR in the two-dimensional case and without loss of generality is

$$\mathbf{K}^T = [1 \; k] \tag{73}$$

$$\Delta\mathbf{A}^T = [\mathbf{A}_{b+f} - \mathbf{A}_b]^T = [d_1, d_2]^T \tag{74}$$

By these definitions, the basis projection angle $\phi = \tan^{-1}(k)$. The optimal basis projection satisfies

$$\left.\frac{\partial\,(\mathrm{SNR})^2}{\partial k}\right|_{k_{\mathrm{opt}}} = 0 \tag{75}$$

There are two solutions. One is

$$k = -d_1/d_2 \tag{76}$$

This is the projection in which the feature vanishes relative to the background. It has zero SNR. The other solution is

$$k_{\mathrm{opt}} = \frac{d_1 \,\mathrm{COV}\,(A_1,A_2) - d_2 \,\mathrm{VAR}\,(A_1)}{d_2 \,\mathrm{COV}\,(A_1,A_2) - d_1 \,\mathrm{VAR}\,(A_2)} \tag{77}$$

This value of k maximizes the SNR; it depends on the system's noise properties and (this is where *a priori* information is essential) on the difference vector **D** of the two substances to be distinguished optimally.

Apparently, the optimal weighting vector $\mathbf{K}_{\mathrm{opt}}$ can be written in matrix notation as

$$\mathbf{K}_{\mathrm{opt}} = R_A^{-1}\,\Delta\mathbf{A} \tag{78}$$

the product of system noise properties multiplied by a vector defining the imaging task. The inverse of the covariance matrix exists when M is nonsingular and when each measurement variance is finite and nonzero.

Equation (78) defines a projection angle which is optimum for a carefully specified imaging task. This projection can be placed within the context of a basis transformation by finding a set of vectors to augment k_{opt} to span the basis space. This set of vectors is not unique, of course. But in two-dimensional space, one pleasing choice is the vector whose projection is uncorrelated with C_{opt}. Let us call this vector **Z** and define q to be the projection which is uncorrelated with C_{opt}:

$$q = \mathbf{Z}^T\mathbf{A} \tag{79}$$

$$\mathrm{COV}\,(C,q) = E\,\{(\Delta C_{\mathrm{opt}})(\Delta q)\} = 0 \tag{80}$$

$$= \mathbf{D}^T\mathbf{R}_A^{-1}\,\mathbf{R}_A\mathbf{Z} = \mathbf{D}^T\mathbf{Z} \tag{81}$$

Therefore, $\mathbf{Z}$ is also the vector which is orthogonal to $\mathbf{D}$. It does not depend on any of the system's noise characteristics since it is the projection which gives zero signal. Lehmann[31] has shown that $\mathbf{Z}$ is also orthogonal to $\mathbf{K}_{\mathrm{opt}}$ for a special definition of orthogonality,

$$\mathbf{K}^T R_A \mathbf{Z} = 0 \qquad \text{for two vectors } \mathbf{K} \text{ and } \mathbf{Z} \tag{82}$$

Therefore, $\mathbf{Z}$ and $\mathbf{K}_{\mathrm{opt}}$ are orthogonal vectors, they span the basis space, and the projections they represent are uncorrelated for a definition of inner product which weights products by the elements of the covariance matrix. The vectors $\mathbf{Z}$ and $\mathbf{K}_{\mathrm{opt}}$ define a new basis set and coordinate frame in which a projection along the one basis vector gives the optimum image. A projection along the other basis vector has zero SNR. It is easily shown that the noise is isotropic within this new coordinate frame. The noise sources along each of the dimensions are uncorrelated, having variance

$$\mathrm{VAR} = d_1^2\,\mathrm{VAR}\,(A_2) + d_2^2\,\mathrm{VAR}\,(A_1) - 2d_2d_1\,\mathrm{COV}\,(A_1,A_2) \tag{83}$$

The optimum value of the SNR is simply

$$\mathrm{SNR}_{\mathrm{opt}} = (\mathbf{D}^T R_A^{-1}\mathbf{D})^{1/2}$$
$$= \left[\frac{d_1^2\,\mathrm{VAR}\,(A_2) + d_2^2\,\mathrm{VAR}\,(A_1) - 2d_1d_2\,\mathrm{COV}\,(A_1,A_2)}{\mathrm{VAR}(A_1)\,\mathrm{VAR}\,(A_2) - \mathrm{COV}^2\,(A_1,A_2)}\right]^{1/2} \tag{84}$$

R_A^{-1} is a positive, definite matrix when the basis functions are not linearly dependent. Therefore, the ratio of signal to random noise at the optimal projection angle is always real and positive (nonzero) whenever D is nonzero.

6.4. Comparison of Noise in Conventional and Energy-Selective Systems

Since energy-selective systems extract more information than conventional systems, it might seem that they should require higher exposure levels. Actually, the opposite is true. This section shows that energy-selective systems extract more information for the same dose than the conventional systems. As in the previous section, noise variance and SNR will be considered separately.

The comparison from the point of view of noise variance is quite simple. It was shown that the optimal synthesized monoenergetic image in a system with Poisson distributed noise has variance equal to

$$\mathrm{VAR}(C) = 1/(N_1 + N_2) \tag{85}$$

This is exactly equal to the noise variance in a conventional system using the same total flux. Thus, there is no increase in noise due to the decomposition process and subsequent recombination to form the synthesized monoenergetic image. Yet, by going to all this trouble, the polyenergetic nonlinearities have been eliminated, the image data are truly system independent (results are repeatable and portable), and the basis set coefficient line integrals are available for material cancellation operations. Thus, from the point of view of variance, the energy-selective system extracts more information for the same dose.

The comparison from the point of view of SNR is more complex. A suitable image task must be found (to define a signal) and an expression for the SNR of conventional systems must be derived for comparison to the energy-selective results [Eq. (84)].

The imaging task is the same as discussed in the previous section on optimal SNR, i.e., distinguish a feature in the presence of a background material (f + b) from the background alone (b). For the comparison to be valid, both systems, energy-selective and conventional, should use the same X-ray technique factors. This will be assured by assuming that both systems use the same measurements. The only difference will be in the processing of data. The conventional system simply sums the measurements and takes the logarithm. The energy-selective system uses the two measurements to calculate the line integrals of the basis set coefficients and then forms the optimal projection. The derivation which follows was first shown by Lehmann.[31]

There are two sets of measurements, $(I_1, I_2)_{\mathrm{b}}$ and $(I_1, I_2)_{\mathrm{b+f}}$, where I_i is the transmitted flux with spectrum i and b denotes a measurement through only the background material, while b + f denotes a measurement through the background and feature combination. Using the definitions of the previous section,

$$A_{1(\mathrm{b+f})} = A_{1\mathrm{b}} + d_1 \tag{86a}$$

$$A_{2(\mathrm{b+f})} = A_{2\mathrm{b}} + d_2 \tag{86b}$$

where d_1 and d_2 are assumed to be small quantities (the feature is thin and the task is challenging). Both the conventional and the energy-selective systems used identical spectra and total exposure. The optimal SNR for the energy-selective system was previously derived in equation (84). Lehmann concluded that the SNR of a conventional system is

$$\mathrm{SNR} = \frac{|(d_1 m_{11} + d_2 m_{12})I_1 + (d_1 m_{21} + d_2 m_{22})I_2|}{(I_1 + I_2)^2} \tag{87}$$

Comparing this result with that of Eq. (84) for the energy-selective case shows that

$$\mathrm{SNR}_{\text{conventional}} \leq \mathrm{SNR}|_{\text{energy-selective}} \tag{88}$$

with equality if and only if

$$(d_1 m_{11} + d_2 m_{12}) = (d_1 m_{21} + d_2 m_{22}) \tag{89}$$

a condition with an interesting physical interpretation. This condition states that the difference in the line integral vectors must be equal for the two measurement spectra. Thus, the energy-selective and conventional systems will have the same SNR only if the effective energies of the two spectra are the same (and the system gathers essentially no energy-dependent information) or the feature has zero attenuation (and there is no signal in either system). In all practical situations, the energy-selective system always has better SNR for identical spectra and radiographic exposures.

6.5. Conspicuity Enhancement

One characteristic of energy-selective generalized projections is that the contrast between any two substances of different effective atomic number is under arbitrary control. Here is a provision for enhancing the relative contrasts of interesting objects while suppressing the contrasts of unimportant features.

For two nonoverlapping substances ξ and ψ, the contrast ratio (CR) of ξ relative to ψ is simply[31]

$$\mathrm{CR}(\phi) = \frac{|\xi| \cos(\theta_\xi - \phi)}{|\psi| \cos(\theta_\psi - \phi)} \tag{90}$$

where $|\xi|$ is the length of the basis vector corresponding to the measurement, θ_ξ is its characteristic angle, and ϕ is the basis projection angle which is a parameter of display. By proper selection of ϕ, the CR can take any real value, negative or positive.

To quantify improvement over single-energy imaging studies, Riederer's[38] "contrast enhancement factor" is adopted. The contrast enhancement factor (CEF), like CR, is a deterministic quantity insensitive to measurement precision but varying with basis projection angle:

$$\text{CEF}(\phi) = \frac{\text{CR}(\phi)}{\text{CR in the conventional image}} \tag{91}$$

If the synthesized monoenergetic image with equivalent angle Ω is used for the conventional images, then

$$\text{CEF}(\phi) = \frac{\cos(\theta_\xi - \phi)\cos(\theta_\psi - \Omega_\text{E})}{\cos(\theta_\psi - \phi)\cos(\theta_\xi - \Omega_\text{E})} \tag{92}$$

When the magnitude of the CEF is less than one, energy selectivity has reduced ξ with respect to ψ. When its magnitude is greater than one, the conspicuity is improved.

Figure 14 shows how the CEF of a 15 mg/cm^2 iodine solution relative to bone and water varies with the basis projection angle. The iodine solution and bone both displace water, the monoenergetic display energy is 50 keV, and the basis units are 0.001 inch Lucite and 0.0001 inch aluminum.

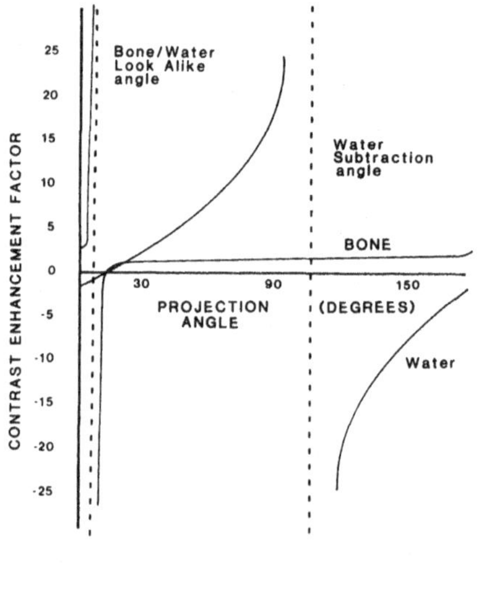

Figure 14. The contrast enhancement factor of a 15 mg/cm^2 iodine solution relative to bone and water at an energy of 50 keV. Basis units are 0.001 inch Lucite and 0.0001 inch aluminum. Where the contrast enhancement factor has a magnitude greater than one, the conspicuity of iodine is improved.

TABLE 2. Contrast Enhancement Factors for Iodine and Bone Displacing Water

Projection	Iodine/water	Iodine/bone
Water cancel	∞	1.3
Bone/water look-alike	−1.3	3
Bone cancel	−0.6	∞

Table 2 lists the CEFs of iodine relative to bone and water for some common projections. Technically, errors in hardening correction, from finite precision arithmetic, and from other sources limit the maximum value of the CEF in a real system to roughly 100.[38]

It is interesting to study the CR of two materials at the cancellation angle of a third material. A common example is the cancellation of water to enhance the relative contrast of iodine. Bone does not cancel; it represents an undesired residue. The CR of iodine to bone at the water cancellation angle is

$$CR = \frac{|I|\,(\tan\theta_W \cos\theta_I - \sin\theta_I)}{|B|\,(\tan\theta_W \cos\theta_B - \sin\theta_B)} \tag{93}$$

With water removed, a 1 g/cm^2 thickness of bone appears equivalent to 5 mg/cm^2 of iodine.

Contrast agents with atomic numbers greater than iodine ($Z = 53$) will be more effective at suppressing undesired bone shadows. Figure 15 demonstrates the dependence of the contrast agent to bone CR on the

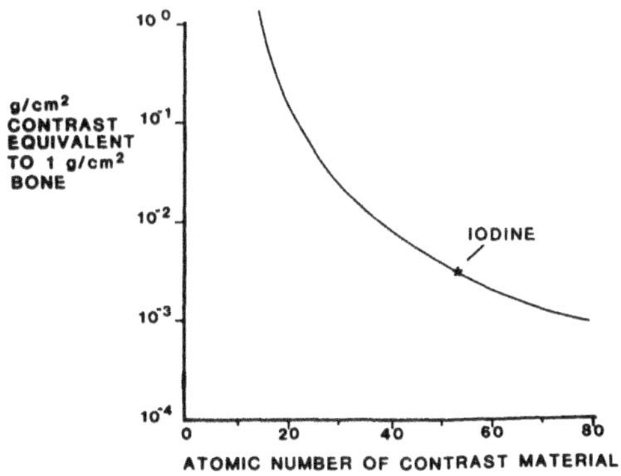

Figure 15. The contrast agent to bone relative contrast as a function of the contrast agent's atomic number, in the water-canceled material selective images.

agent's atomic number. Contrast agents with high atomic number could possibly reduce the relative bone residual tenfold.

7. CONCLUSION

The energy spectrum of X-rays transmitted through the body contains a great deal of information which can be used to reduce the effects of two important sources of noise in diagnostic imaging: overlying anatomical detail and quantum random noise. Energy-selective data make possible CT reconstructions in which there are no polyenergetic nonlinearities and artifacts. In any properly constructed energy-selective system, CT or single-projection, the resulting images are truly system independent and portable. As shown, the energy-selective results are much more amenable to quantification, analysis, and standardization. Energy selectivity provides capability which has been lacking in single-projection radiography, that of identifying the approximate "effective atomic number" of an otherwise unknown feature.

Energy-dependent techniques actually reduce the effects of X-ray quantum noise. The systems extract more information for the same radiographic exposure than do conventional systems. Furthermore, they can detect small features with SNRs better than conventional systems utilizing identical exposure factors.

There is a certain inherent flexibility in the energy-selective measurements. They contain all the information of the object using X-ray measurements which can be inferred, and thus contain more information than can be displayed at one time. Only one set of patient transmission measurements suffices to produce a wide range of image types which can be selected and displayed after the time of patient exposure. The viewer can choose a bone-canceled image, and then later change his mind and select a water-canceled image for viewing without the need for reexposure. At the time of viewing, the contrast between any two materials of different "effective atomic number" is an arbitrary parameter.

From the point of view of either anatomical noise or quantum noise, energy-selective systems extract information more efficiently than conventional systems. This is due to the use of *a priori* knowledge of the physics of X-ray interactions with matter expressed using a vector space description of the energy dependence of these interactions. This description is sufficiently accurate for even the most quantitative applications, yet it reflects the fundamental simplicity of the physics and allows information to be extracted with apparatus which is both useful and practical.

REFERENCES

1. B. Jacobson, Dichromography—A method for in-vivo quantitative analysis of certain elements, *Science* **128,** 1348 (1958).
2. C. A. Mistrette, M. G. Ort, F. Kelcz, J. R. Cameron, M. P. Sieband, and A. B. Crummy, Absorption edge fluoroscopy using quasimonoenergetic x-ray beams, *Invest. Radiol.* **8,** 402 (1973).
3. F. Kelcz, C. A. Mistrette, and S. J. Riederer, Special considerations for absorption edge fluoroscopy, *Med. Phys.* **4** (1977).
4. R. A. Kruger, C. A. Mistretta, A. B. Crummy, J. F. Sackett, M. Goodsitt, S. J. Riederer, T. L. Houk, C.-G. Shaw, and D. Flemming, Digital k-edge subtraction radiography, *Radiology* **125,** 243–245 (1977).
5. F. Kelcz and C. A. Mistretta, Absorption edge fluoroscopy using a three spectrum technique, *Med. Phys.* **3,** (1976).
6. P. S. Yeh, Selective material imaging using multiple energy measurements, Ph.D. dissertation, Department of Electrical Engineering, Stanford University, Stanford, California (1980).
7. P. S. Yeh, A. Macovski, and W. Brody, Noise analysis in isolation of iodine using three energies, *Med. Phys.* **7,** 634–643 (1980).
8. A. Macovski, R. E. Alvarez, L. A. Lehmann, E. Roth, and W. R. Brody, Iodine imaging using three spectra, *Proc. SPIE* **314,** 140 (1981).
9. A. Macovski, R. E. Alvarez, and J. Chan, Selective material x-ray imaging using spatial frequency multiplexing, *Appl. Opt.* **13,** 2202 (1974).
10. A. Macovski, R. E. Alvarez, and J. Chan, Spectral selectivity in radiography using a grating encoding system, *Proc. SPIE* **52** (1975).
11. R. E. Alvarez, Extraction of energy dependent information in radiography, Ph.D. dissertation, Department of Electrical Engineering, Stanford University (1976).
12. E. Rubenstein, et al., Synchrotron radiation and its application to digital subtraction angiography, *Proc. SPIE* **314,** 42 (1981).
13. G. E. Donovan and G. Jones, Colour in radiography, *Proc. R. Soc. Med.* **44,** 816–818 (1951).
14. S. Takahashi, Chromatoroentgenography: A method of taking the colored roentgenogram on the multilayer color film, *Tohoku J. Exp. Med.* **56,** 43–45 (1952).
15. B. Jacobson and R. S. MacKay, Radiological contrast enhancement methods, *Adv. Biol. Med. Phys.* **6,** 201–261 (1958).
16. A. Hall, N. Pelc, S. J. Riederer, G. S. Keyes, W. R. Brody, L. A. Lehmann, A. Mackovski, and R. E. Alvarez, An experimental system for dual energy scanned projection radiography, *Proceedings of SPIE, 314, Conference on Digital Radiography, 155* (1981).
17. F. G. Sommer, W. R. Brody, D. Gross, A. Macovski, A. Hall, and N. Pelc, Excretory urography using dual-energy scanned projection radiography, *Radiology* **141,** 529–532 (1981).
18. F. G. Sommer, W. R. Brody, D. Gross, A. Macovski, A. Hall, and N. Pelc, Dual energy scanned projection radiography, *Appl. Radiol.* **2,** 59–66 (1982).
19. R. A. Kruger, J. D. Armstrong, J. A. Sorenson, and L. T. Niklason, Dual energy film subtraction technique for detecting calcification in solitary pulmonary nodules, *Radiology* **140,** 213–219 (1981).
20. R. T. Ritchings and B. R. Pullan, A technique for simultaneous dual energy scanning, *J. Comput. Assist. Tomogr.* **3, 842–846 (1979).**

21. R. A. Brooks and G. DiChiro, Split-detector computed tomography: A preliminary report, *Radiology* **126,** 255–257 (1978).
22. A. Fenster, Split xenon detector for tomochemistry in computed tomography, *J. Comput. Assist. Tomogr.* **2,** 243–252 (1978).
23. R. T. Stone, An energy discriminating detection technique for medical X-ray imaging, Ph.D. dissertation, Department of Electrical Engineering, Stanford University (1981).
24. B. Strul, Energy spectral analysis in projection radiography, Ph.D. dissertation, Department of Electrical Engineering, Stanford University (1981).
25. R. A. Rutherford, B. R. Pullan, and I. Isherwood, X-ray energies for effective atomic number determination, *Neuroradiology* **11,** 23–28 (1976).
26. R. A. Rutherford, B. R. Pullan, and I. Isherwood, Measurement of effective atomic number and electron density using an EMI scanner, *Neuroradiology* **11,** 15–21 (1976).
27. E. C. McCullough, Photon attenuation in computed tomography, *Med. Phys.* **2,** 307–320 (1975).
28. B. K. Agarwal, *X-ray Spectroscopy,* Springer-Verlag, Berlin (1979).
29. R. E. Alvarez, Energy dependent information in X-ray imaging. Part 1. The vector space description, in press.
30. R. E. Alvarez and E. Seppi, Comparison of noise and dose in conventional and energy selective tomography, *IEEE Trans. Nucl. Sci.* **NS-26,** 2853–2856 (1979).
31. L. A. Lehmann, Utilization of multi-spectral measurements in radiography, Ph.D. dissertation, Department of Electrical Engineering, Stanford University (1982).
32. J. Heinzerling and M. Schlinwein, Nonlinear techniques in multispectral x-ray imaging, *IEEE Trans. Nucl. Sci.* **NS-27,** 961–967 (1980).
33. L. A. Lehmann, R. E. Alvarez, W. R. Brody, N. Pelc, S. J. Riederer, and A. L. Hall, Generalized image combination in dual kVp digital radiography, *Med. Phys.* **8,** 659–667 (1981).
34. B. Rutt and A. Fenster, Split-filter computed tomography: A simple technique for dual energy scanning, *J. Comput. Assist. Tomogr.* **4,** 501–509 (1980).
35. D. J. Drost and A. Fenster, Experimental dual xenon detectors for quantitative CT and spectral artifact correction, *Med. Phys.* **7,** 101–107 (1980).
36. E. Parzen, *Stochastic Process,* Holden–Day, San Francisco (1982).
37. R. E. Alvarez and L. A. Lehmann, Energy dependent information in X-ray imaging. Part 2. Information extraction and noise, in press.
38. S. J. Riederer, R. A. Kruger, and C. A. Mistretta, Limitations to iodine isolation imaging using a dual beam non-k-edge approach, *Med. Phys.* **8**(1), 54–61 (1981).

Index

GPSR Compliance

The European Union's (EU) General Product Safety Regulation (GPSR) is a set of rules that requires consumer products to be safe and our obligations to ensure this.

If you have any concerns about our products, you can contact us on ProductSafety@springernature.com

In case Publisher is established outside the EU, the EU authorized representative is:

Springer Nature Customer Service Center GmbH
Europaplatz 3
69115 Heidelberg, Germany

Batch number: 10854084

Printed by Printforce, the Netherlands